I0467828

JUNK

YARD

MARS

Copyright © 2014 by N.B. McKinney

All rights reserved, including the right to reproduce this book or portions thereof in any form whatsoever. For permission to reproduce selections from this book, or to contact the author, write to:

JunkyardMarsBook@gmail.com

Junkyard Mars
Library of Congress Control Number: 2014915105

ISBN-13: 978-1500270247
ISBN-10: 1500270245

Published in the United States of America

First Edition

Borne of my love of Earth, and even greater love of humankind.

Contents

The Freak-Out book of the decade.

"Breathtaking in its scope, heart stopping in its clarity, earthshaking in its relevance."

The thrilling exposé of what Mars is really like – and what NASA won't tell you.

Do not read this book cover to cover or you will overdose on reality.

"An instant classic. If it doesn't amuse you, it will infuriate you."

"The 'make sense or drop dead' book of the century."

"Too smart for its own good."

"A stunning breakthrough."

"Simply brilliant."

"There is nothing in the desert, and no man needs nothing."

T. E. Lawrence
Lawrence of Arabia

Preface

‖‖

Whether or not we ever get to Mars seems less a matter of scientific progress than the balance of power between sane and crazy.

It's interesting that rocket scientists, astronomers, and others in the space industry would be taking the lead in colonizing Mars, because living on Mars would have nothing to do with any of these things.

LET'S FACE IT, NOBODY REALLY WANTS TO LIVE ON MARS. Sure, there may be a certain thrill, and prestige, in being involved with "something big," but...nobody really wants to live on Mars! Then why do we have an ever increasing infrastructure dedicated to what could be the most glorious fraud of all time, involving governments, universities, and private corporations? Why are the careers and futures of so many propped up on dreams and fantasies? This is busy-work taken to its most extreme, another example of the hive mind – and all in the name of...science! Has the entire world gone mad? Join me as I explore this madness – and hope it's not too late.

Whether or not we can get to and live on Mars isn't a simple matter of optimism or pessimism, where optimism says we can and pessimism says we can't, in which case many will accuse me of being a pessimist. It may be necessary to rethink what these two things mean in this new context, in order to prove to you that...I am actually an optimist! Regarding the future of Earth, of course! Which is, after all, what this is about. Even something simple like getting to the top of Everest isn't a matter of optimism. All these things have nothing to do with optimism or point of view. Rather, it all comes down to knowledge and understanding of the basic facts. And the facts are in, they are known. Mars is no longer the mystery it once was.

If there was a single spark that prompted this book, it is the declaration made by the Mars advocates that going to Mars is vital. World hunger and starvation, overpopulation, collapse of economies (or threat of collapse), unemployment (which is more than 25% in some age groups), America at war (with one country or another, take your pick), civil wars and unrest in the Mideast, drug resistant bacterial strains, unstoppable tuition inflation, maintenance of our nations bridges (thousands of which are almost beyond repair), and so on ad infinitum...given this spectrum of problems, if the skills, abilities, and foresight of the engineering community are so transcendent, why do the same old problems continue to be the same old problems? Our noble engineers obviously have an extremely misguided sense of ethics, mission, and timing if they think that solutions to these problems aren't vital but – of all things – getting to Mars is!

Not only does the mainstream think that engineers are magicians, this seems to be fostered by the engineering community itself, and as a form of magna carta. It's the modern version of alchemy and a colony on Mars would require nothing less than this. And whereas the ancient alchemists believed they could turn lead into gold – which is of course impossible – today's engineers – who might not deny that it would take alchemy – think they can perform the alchemy of putting a colony on Mars. And whereas the mainstream doesn't just expect this from engineers, engineers have accepted this burden, but furthermore seem to bear it as a duty. It's one thing to be at the top of a hierarchy – and somehow engineers have become this – but having accepted the role of the alchemist, still, they are not beyond the reach of criticism or scrutiny. But you would never know that, given the media's intrigue with outer space and Mars – while their continued intellectual passivity, happy-go-lucky reporting, and careless blending of science fact and fiction does nothing but lead to an irrational exuberance for outer space and Mars, and brainwash and provoke the public – many in the media are themselves victims of the brainwashing.

The media's complete lack of a balanced skepticism is outrageous.

Rather than offer a much needed counterpoint, they seem hell-bent as the voice for the propaganda, emanating from NASA and others in that arena.

Since there doesn't already seem to be one, let this be considered a proper and complete anti-thesis, and for it to be considered timeless, as relevant one thousand generations from now as it is today, regarding the question of whether or not we go to, and colonize, Mars. Or might this be considered the thesis, since the question of whether or not we go to, and colonize, Mars hasn't, or so it would seem, itself already been formalized as a proper thesis – but rather has been a series of stumbles, daredevil stunts, and crackpot ideas originating in the dark recesses of history, and our imaginations. So – join me as I attempt to redeem planet Earth – for the sake of Humanity.

Author's Note

Pop cultural references to books and movies are provided as a convenience to the reader. They are for illustrative purposes only and were not a source of information or inspiration. Scientific achievement and the experience of being human are my sole source of inspiration.

Introduction

We're warm blooded! It's incongruous that we would live on a planet that is entirely frozen – and to think this is an inevitable choice or destiny.

Even if we discovered that the seeds of our existence on Earth somehow originated on Mars – then got here via a long and cold ride through the vacuum of space on a chunk of debris from some cosmic collision – it seems the investment Earth has made in us since then has precluded the possibility that a successor should come easy – and vice versa: the investment we've made in being earthlings.

IT'S AMAZING THAT ANY SCIENTIST COULD THINK, TODAY, given what we know, that we could live on Mars, or cultivate plant crops...to actually colonize – colonize! – Mars. Even before the rovers – before satellites – before anything that is modern – only the most imaginative would be taken by such a fantasy! But now? The lessons learned from the rovers are already old! It's as if we've gone retrograde, and decided to just IGNORE the science! All the dumb, crazy scientists creating dumb, crazy science, thinking they're saving humanity while America and the rest of the world goes to hell in a handbasket...which is happening because of the dumb crazy science cycle, or the **smart science-crazy science cycle**. The scientists are creating the very problems we need to escape from – are they all just nincompoops like the rest of us?

MANY EXAMPLES OF LEGITIMATE SCIENCE HAVE TURNED into junk, and the junk science is incentivized by all the money thrown at it, resulting in a feedback loop that has become a monster, all of which involves countless and ridiculous claims, promises, predictions, and misperceptions about reality and our powers to control it – while at the same time we're increasingly able to realize what we had always only been able to imagine and the border between science and science-fiction becomes a double-edged sword. But for the sake of humanity, the exploration of space needs to be guided by hard-edged sanity, not delusional hopes and dreams (or psychos).

OUR CONTINUED FASCINATION WITH MARS DEMONSTRATES, if nothing else, how we hang onto both the past, and to antiquated visions of the future. That's right – our vision of man on Mars is not futuristic, but antiquated.

Mars – is passé. Manned space flight was a phase that has run its course, and henceforward is a complete waste of time – or a frivolous diversion of the idle rich, which it is quickly becoming.

There are, to be sure, deeply fundamental problems with the idea of living creatures from one planet living on another. In science-fiction this may be commonplace, but reality... has rules. We are specifically designed to live in a particular ecosystem – through evolution and adaptation.

THAT MAN WILL "JUST EVOLVE" IS ACTUALLY ONE OF THE more mind-blowing assumptions made by the Mars advocates, since it is more likely that, on Mars, man will "just die." Evolution allows for possibilities, not certainties. Evolution increases the odds that something will survive – but not necessarily us. Years after we attempt a colony on Mars, it may be home to numerous living species – but with no trace of humans whatsoever!

LIFE ITSELF IS AN ANOMALY, THE ODDS AGAINST OUR EXIST-ing at all are incomprehensible, but somehow life has sprung forth, and manages to persist – but through a form of compromise, in which we, our bodies, are in a constant state of deterioration; nature fully intends and conspires to reclaim us, and man is in a constant struggle to combat this deterioration. We've invented many ways to do this, name-ly, the machinery of our lives. But none of this machinery will exist on Mars! This means that the natural deterioration we experience on Earth would, on Mars, only be worse, in a steady state of FREE FALL – unabated – unstoppable.

IN DEFYING THE PHYSICS THAT KEEPS US HERE, IT MAY BE IM-possible for us not to destroy ourselves – what grand creatures we must be to think we should. It may be a great-er freedom to simply embrace our mortality, and the relative tranquility of what we already know.

♦ ♦ ♦ ♦ ♦

What will we be eating on Mars? What will our days be like? Will it really be as adventurous as Elon Musk of SpaceX is promoting?

You won't be able to bring your favorite things with you, things you might bring on a car or plane trip – and not be-cause these things wouldn't fit on board a spaceship.

Give up all my favorite things? I couldn't do that. Includ-ing the dog? Football? What about the kids? Will my kids be crying a lot as the launch date – and the moment of truth – draws near? Will I never have sex with my wife again? It is inconceivable that all this would happen without a large part of me dying. There is simply no way to compensate for this enormous loss by way of some abstract sense of thrill or adventure. Even Christopher Columbus returned home. Even the most extreme, the most daring adventurer

has always returned home, right up through the lunar astronauts, the Space Shuttle, and the ISS. All these missions
have always had as their ultimate objective a safe return
home – and which seems axiomatic, itself a force of nature.
To forsake this instinct, this undeniable force of nature, we
are going in a different direction entirely, a direction no man
or beast has ever gone. This is not just vague apprehension
toward the unknown. Nor can a colony be explained or
rationalized simply as adventure, exploration, or curiosity.

◆ ◆ ◆ ◆ ◆

The case I'm making against Mars isn't based on a lack
of or deficiencies in rocket technology. We have rockets or
spacecraft that can get to Mars, and beyond, and it's not
a big assumption that improvements can be made in this
area, even in the short term (10-50 years), so none of my
case against Mars depends on such argumentation, and I
consider myself free of that burden – although I will make
incidental references to it since it's impossible to discuss a
colonization of Mars otherwise. Whether the trip takes six,
seven, or eight months is also not the issue – that is more
a concern of the Mars advocates than the skeptics. The
rocket science is actually the easy part although it will take
the ingenuity of the Wright brothers or Edison to overcome
some remaining rocket issues – as in carrying live humans,
and heavy payloads are impossible to land gently as of this
writing, and there's no technology that would allow for a
return trip. Plus they're slow. Each of these presents enormous challenges that must be overcome. The once every
twenty-six month optimum launch window is an even bigger obstacle and nothing can be done about this. But the
rocket science constitutes only about 1% of the challenge
posed by a self-sustaining colony. It's the other 99% that we
consider now: humans actually living on Mars.

THIS IS IMMINENTLY MORE FEASIBLE THAN A MARTIAN EARTH SUBSTITUTE: THAT AS WE'RE MAKING HUGE SACRIFICES TO CREATE THIS MARTIAN EARTH SUBSTITUTE, TODAY'S ISSUES ARE RESOLVED, AND LIFE ON EARTH GETS BETTER – A STABLE WORLD PEACE IS ESTABLISHED AND OUR RELIANCE UPON OIL IS REPLACED WITH OTHER TECHNOLOGIES, AND OTHER WONDERFUL TECHNOLOGIES EMERGE. OUR PRESENCE ON MARS WILL THEN SEEM LIKE ORWELLIAN FOOLISHNESS, AT LEAST TO THOSE ON MARS GAZING UPON EARTH WITH A NEWFOUND ENVY, PAWNS OF A GIGANTIC BRAINWASHING, OVERAMBITIOUS RUBES CAUGHT IN A COSMIC RUBICON.

2

Mirror Image Earth

Will I grow old on Mars, or will Mars make me grow old?

So you think we should put a colony on Mars. You may even want to go to Mars yourself. There are some adversities you'll need to consider, adversities that NASA and the Mars advocates aren't exactly proud of and are doing their best to keep secret. Let me tell you about the Mars NASA doesn't want you to know about. The dead and deadly Mars.

Even if Mars was a MIRROR IMAGE of Earth (minus humans or the results of human activity), there would be nothing there, of course, starting out. There would be water, an oxygen-rich atmosphere, and gravity like on Earth, but no infrastructure, no transportation, no bridges, no running water, no electricity, no accommodations. A lot of the challenges that man would face on Mars would also exist on such an imaginary mirror image planet, where everything would seem to be on the plus side. But a mirror image of Earth would be like an Amazon jungle or African game preserve – a zoo – where you'd be more likely eaten alive than anything else. This would be an improvement over Mars, of course, but not by much. *This* would be a world fit for adventure seekers! A survivalist's dream come true. Such a world might be downright comfortable for a primitive tribe in the Amazon, not a bunch of teched-out venture capitalists led by Elon Musk. There would still remain, of course, the huge problem of getting there. Now switch to Mars – it is impossible even to imagine this same teched-out bunch...on the dead and deadly Mars! This could not possibly ever happen, not in a million years! Yet space tourism promoters seem to think that Mars is just that, a mirror image of Earth.

That would be a great premise for a science-fiction novel – our arrival on a mirror image of planet Earth. We'd touch down near a large body of fresh water, of course, where the climate was hospitable year round, because you wouldn't want to immediately be challenged by cold, much less freezing weather nor, for that matter, a desert environment. We'd need every conceivable advantage, and should not be handicapped in any way whatsoever, so the landing spot would require much consideration. Don't give me that heroic "hard and hearty" *adaptation* bullshit – you'd rather step out into a foot of snow on a planet defined by an *endless winter or freezing carbon dioxide showers*? What kind of nightmare is that? It better be a freaking *Garden of Eden*. After being cooped up on a spaceship for several, perhaps many years, like a dog in a cage, *we would need to land on a mirror image of planet Earth, of course!* And we really do not have many choices regarding this! Because even in a Garden of Eden you're going to have spiders and snakes. Anyway, back to our mirror image of Earth premise, the area equivalent to southern California might satisfy these conditions, which provides expansive open areas for landing many spaceships, is somewhat desert but has some natural lakes – in contrast to the area equivalent to the eastern coast of the United States, which is of course more watery but which would be heavily and uniformly forested and has severe winters. But there are many other places to choose from worldwide, there's no reason to have a bias starting out toward what is equivalent to North America or the United States, all of which might be considered too cold generally. The first settlement should be near an ocean, to support a diet that includes fish and other seafood. There wouldn't really need to be a master plan – one doesn't need the cunning of Machiavelli – just live in harmony with nature and each other! Things would be very simple at first, primitive, perhaps becoming equivalent to medieval life in only a few generations since we'd have somewhat of a head start given our twenty-first century knowledge.

We would, ultimately, be starting from scratch even on a mirror image of planet Earth, which would be many light-years away so we couldn't just ship anything and everything – perhaps just you and the shirt on your back. And who's to say that progress should be driven the same way as on Earth, or mechanized, or dominated by technology? Must everything be recreated as on Earth? Right down to the last cell-phone tower? This would *not* be the reasoning behind a mirror image Earth – that's your earth-brain thinking again. Perhaps the next time around we would be better off living in some primal state.

But the Mars advocates think we can do this on Mars? *The Mars advocates think we can do this on Mars!*

3

You Like to Breathe, Don't You?

*"There is nothing in the desert,
and no man needs nothing."*

T. E. Lawrence
Lawrence of Arabia

**Man can't live on Mars, other than as a ghastly,
weird experiment, like pulling the wings off a
fly while proclaiming "look, it's still alive!" in the
moments before it dies.**

SO FAR I'VE MADE TWO VERY IMPORTANT POINTS, WHICH SHOULD
clarify some misperceptions: that the rocket science isn't the main
obstacle preventing a colony on Mars, and that extreme challenges concerning basic habitability would be experienced even if Mars was like
Earth. Let's now examine the issue of habitability more closely, and the
idea of not simply visiting, but of colonizing Mars.

It has been discovered, in recent decades, through much evidence
collected through satellite reconnaissance and *four* NASA rovers, that
Mars is very different from Earth, so much so that it can not be considered either habitable or hospitable and should be considered to be
quite deadly, with temperatures much lower than anywhere on Earth
except for the poles, an atmosphere that is less than 1% that of Earth's in

density and that consists of 95% carbon dioxide and almost no oxygen, extreme amounts of solar radiation due to the thin atmosphere and very weak magnetosphere, no liquid water, and no vegetation whatsoever and therefore no natural food source. This means that a colony might be considered "doable," and for Mars to be considered "habitable," only if multiple and highly complex life support measures were taken and maintained, including living underground, or above ground in habitats that provide an entirely artificial living environment. One could only live on Mars if in a "Mars-proof" habitat – but certainly this is just an ironic way of saying that Mars is *not habitable!* One could counter this by saying that our homes and buildings are nothing but "Earth-proof" habitats, but this ignores the extreme and unearthly aspects of Mars, besides the fact that living creatures exist in harmony with their environment, not separate from it. Human life in particular extends way beyond the confines of the structures we build – we are more then pets confined to a cage, or potted plants sitting on a window sill.

It takes about 8 months for unmanned rockets to get to Mars – that's the extent of the Mars program as of the beginning of 2013. NASA and a few other countries have been sending rockets and probes to Mars since 1960, all of them unmanned but carrying either satellites or, more recently, "dune buggies" to roll around on the surface, like robots of course since we don't send people up with them, and which are remotely controlled from Earth. The dune buggies, of which there have been four, called rovers, carry sophisticated instruments including lasers, cameras, and analyzers, and which, among other things, take small samples of the Martian dirt and air, which it tests and analyzes right there. Pause – all of which is, of course, freaking amazing. The rovers are designed to communicate with Earth and send back data concerning the various soil and air analyses. In this way we've been able to determine, *with great precision*, the exact nature of Martian air and soil, at least in the samples that have been tested, and which are so numerous as to be countless.

So we know what's up there – and combined with high-definition satellite imagery of the Martian landscape that we've been accumulating for years...we know a lot. We have the empirical evidence. More than enough to know that the prospect of man living on Mars basically...isn't doable – unless of course you're a Mars Advocate – and believe we should not only send people to Mars, but colonize it! Which is crazy because, in a nutshell, Mars is a subfreezing desert with almost no oxygen and no liquid water – and no signs of life whatsoever, even at the microscopic level. "Organic molecules" detected in soil samples are provocative and equivocal to some but a preoccupation with this degree of detail or scrutiny demonstrates a rejection of the big picture – in high-definition and 3D – which is preposterously misguided, since whatever microscopic details are largely irrelevant to the long term prospects of a human colony.

Here are the cold, hard facts (and they really are cold!).

WATER

There is no liquid water anywhere on the entire planet. There is, however, water at both the north and south poles permanently locked up as ice – polar ice caps, similar to Earth's north and south poles – and elsewhere in scant amounts frozen into the Martian ground, also permanently locked up. Radar measurements of the north polar ice cap found the volume of water ice in the cap to be 821,000 cubic kilometers (197,000 cubic miles). That's equal to 30% of the Earth's Greenland ice sheet. The radar is onboard the Mars Reconnaissance Orbiter [Wikipedia, Martian polar ice caps]. None of this water is in a readily useable state, especially whatever "moisture" is frozen in the ground. In fact there is a layer of frozen carbon dioxide on top of the water ice at both poles, which sublimates entirely from the northern pole each summer (unlike the southern pole), with the frozen water ice cap remaining, this cycle repeating annually, but which makes this (otherwise convenient?) water-ice relatively inaccessible much of the time.

> **Based on all the evidence, whatever water that exists on Mars is either in a solid state or a vapor state – but never as a liquid.**
>
> **Even here on Earth, the water planet, there are shortages of water, where it imposes great limitations on our behavior and activities. This could only be many times worse on Mars.**

It wouldn't matter how much water exists in the Martian air as vapor. Humans consume HUGE amounts of water, especially the more advanced sections of our society. Can you imagine relying solely on water extracted from the atmosphere? Not in the volumes we require to live, and especially not if you expect any "industry" on Mars. That would require heavy machinery, on an industrial scale, a scenario that is also completely contradicted by the primitive state any colony is likely to assume in the short to mid term. Nowhere on Earth do we resort to such measures, as well as we might, given our soggy, humid atmosphere.

Mars advocates take for granted the role that water plays in not just our lives, but our evolution, in the development of cities and towns. Humans don't just need water to drink. We need oceans and rivers of it. We are water creatures. That's why our desert states aren't just sparsely populated, they're virtually empty! Every single major and minor city on Earth is adjacent to one or more large bodies of water. Obviously a city can't develop without this proximity to water. There is not a single town or city on planet Earth that developed without convenient access to abundant amounts of water – clear, sparkling, delicious water. We could not survive on a desert planet, much less thrive, and no amount of technology, or wishful thinking, can convert Mars into a water planet.

There may be a fundamental problem with creatures from one planet living on another, insofar as its constituent parts are – more than we might ever know – a function of the planet that gave rise to it.

TEMPERATURES

The unrelenting cold on Mars can't be overstated. The entire planet is a subfreezing desert and there is no relief from this. It is a cold, cold, cold planet! There are no warm spots. Think of a planet sized Antarctica with reddish dirt rather than snow (besides all the other problems). Do not be misled by the fact that the temps occasionally get above freezing. Do not be misled by "record highs" that may be 70 F and above, by the equator. The cold and desolation alone make Mars the last place anyone would want to be and completely unfit for humans *or any other living thing*.

What's the temperature on Mars? Well, what's the temperature on Earth? It depends on where and when, and averages and record highs and lows can be misleading, since the temperature at any given time and place may not ever be the average. But Mars is definitely cold, everywhere. Dangerously cold, and sometimes "instant death" cold. The temperatures swing wildly during the day and night and over the surface of the planet, and most statistics regarding this describe only extreme highs, extreme lows, and averages, rather than the "typical daytime high" as might be useful to someone living on Mars or planning to move there. As on Earth, the statistical extremes do not represent the typical, and can be misleading or even useless.

Generally, exposing unprotected skin to "typical" daytime temps would "probably" result in flash frostbite and (in combination with the extremely low air pressure) spontaneous ruptures of skin blood vessels – meaning you'd explode – if not like a stick of TNT.

One source (Quest.NASA.gov) states an average daytime temperature of minus 58 degrees F in the mid-latitudes, with a nighttime minimum of minus 76 F. But like on Earth, it gets considerably colder toward the poles in winter (minus 225 degrees F although this may not be typical) and considerably warmer toward the equator in the summer (95 degrees F although this may not be typical).

Another source claims the two Viking landers recorded temps of 1 degree F to minus 178 degrees F (http://www.astronomycafe.net/qadir/q2681.html).

Now wait a second – Mars has a summer? Because of the polar axis tilt and the elliptical orbit of the planet, Mars experiences what might be called, for lack of a better word, seasons. This adds to the temperature extremes, but think summer in Antarctica rather than summer at the beach. But even this analogy somehow fails because, due to the lack of an atmosphere to keep the temperatures stable,

hourly fluctuations in temperature, through the course of each and every day, are so extreme that the time of day would be more relevant to human activity and productivity on Mars than the time of year or season.

Many of these extreme temperatures *don't occur on Earth* – other than Earth's north and south poles and which are considered uninhabitable for this reason. Clothing alone would not be sufficient to stay warm; heating elements would need to be built into the spacesuit, like an electric blanket.

The word summer has unmistakable and deeply rooted connotations that, when speaking of Mars, can only mislead, and severely.

The rover Curiosity's (in operation since September 2012) onboard weather station measured daytime air temperatures at Gale Crater at just above freezing on many days, including as high as 43 degrees F.

Air temperatures at the Martian surface drop dramatically after the sun goes down, plunging as low as 94 degrees below zero Fahrenheit just before dawn in some places – which is of course deadly. Such big swings in temperature occur because of the effects of solar heating on the extremely thin atmosphere and the dry surface of the planet.

I stated earlier there is a layer – several meters thick – of carbon dioxide ice on top of the water ice caps at both poles. This demonstrates how cold Mars is! It only needs to be 32 degrees F for water to freeze (on both Mars and Earth), but it needs to be much colder for carbon dioxide to freeze, which on Mars is minus 184 degrees F, and which is of course the temp at the Martian poles, and colder (year round at the southern pole as evidenced by the perpetual layer of frozen carbon dioxide). Above that temperature carbon dioxide sublimates (goes from a solid to a gas state) and, on Mars as on Earth, is never found in a liquid state. Carbon dioxide freezes at the much warmer minus 108 F on Earth because of the vast difference in air pressure of the two planets. [http://www.esa.int/Our_Activities/Space_Science/Mars_Express/Mars_polar_cap_mystery_solved].

Don't assume that if you move to Mars you could just live at the equator, since you would still need to wear a spacesuit ALL THE TIME because of the other extremes which you would need to avoid entirely. And spacesuits seem to be inherently better suited to colder weather.

A spacesuit with only heating elements – a simpler design being much preferred over one also requiring cooling elements – would lend itself to living in the cooler mid-latitudes. Considering that you could never take off the spacesuit (while outdoors), if the heating elements didn't work the spacesuit would still provide some insulating warmth – in combination with a parka, perhaps – whereas if the cooling system didn't work and you were living in a warmer climate, towards the equator, on a particularly warm day...you'd be in real trouble.

So in choosing a site for a colony there will be a struggle to find a compromise between regions that may be warmer, or proximity to larger reserves of frozen water – which are confined to the much colder polar regions.

The cold temps alone may not make Mars unlivable, and may not be the strongest argument to turn away from Mars – they may be acceptable for a scientific research camp, but are certainly depressingly inconvenient otherwise. And whether it's plus 34 F or minus 50 F during the day is a moot distinction because I wouldn't be able to run around and play Frisbee in either one because of the other deadly environmental factors for which I would need a spacesuit. Nobody wants to live like an Eskimo. Imagine never being able to enjoy a nice spring or summer day – that seems like a large part of the human experience. Certainly one could never have a nice day if they were always in a spacesuit! Some would argue this by saying that we would simply adapt to wearing spacesuits, such that even though enjoyment of the weather is subjective, people on Mars would enjoy the weather on Mars as people on Earth enjoy the weather on Earth. But our enjoyment of the weather, or environment generally, isn't just a state of mind, but is defined by the effort necessary to protect ourselves from it, through clothing and the *machinery of our lives*. And since man on Mars would *always* be making considerable and time-consuming efforts to protect himself from the environment, far in excess of the rest of us – in the form of spacesuits and other contrivances that are in fact not a natural part of the human experience...this seems to put a considerably lower limit on his ability to enjoy the environment in a very literal and not subjective sense.

But what is the lower limit of cold, that man can tolerate even while second skinning himself with a personal force field of heat? The extremely cold temperatures may be the one challenge that, by itself, could be engineered away – but just barely. It seems we might go from "unlivable" to "do-able" in a single step, but where is that point on the thermometer? We are, after all, warm-blooded. A cold-blooded adaptation might solve this problem, while also lowering the requirements of a high-calorie diet. But that might also be a useful adaptation to Earth – for the same reasons.

AIR PRESSURE AND OXYGEN

No Air – No Oxygen

If a dollar bill represents the air pressure on Earth, then just over half a penny, or six-tenths of a penny, represents the average air pressure on Mars (1,013 millibars mercury on Earth, 6.0 millibars on Mars). This means man can not live on Mars – at least not without a pressure suit, but this would not allow us to preserve "life as we know it."

This example demonstrates how different things are on Mars: if someone standing out in the open on Mars poured some water heated to, say, 70 degrees F, into a coffee cup, right before their eyes it would begin to boil away – perhaps explosively (because the vapor pressure of the water at 70 degrees would far exceed the air pressure) and be gone even if the air temperature was well below freezing, in which case it would boil away before it got a chance to freeze. You wouldn't have a cup of freezing water, but an empty cup. That's how strange things are on Mars.

The reason we're very unlikely to find liquid water on Mars isn't because it's always freezing on Mars – because it isn't always freezing on Mars. The reason we're very unlikely to find liquid water on Mars is because the air pressure is so low! But what does air pressure have to do with liquid water? Everything!

THE TRIPLE-POINT

This leads to an interesting argument made by the Mars advocates, who point out a property of water known as its triple-point, where, at a particular temperature and air pressure, water can coexist as a solid, a liquid, and a gas (which are the three basic states of matter, hence triple-point). Fancy that! The triple-point of water does not occur on Earth, where this property has meaning only to scientists, but it does, at certain times and places, occur on Mars.[1] Double fancy that! In fact the triple-point pressure of water (6.1173 millibars mercury) practically coincides with the average air pressure on Mars – which seems like an amazing coincidence (and bearing in mind the triple-point of water is the same on both Earth, Mars, and everywhere else in the universe). But that does not mean that liquid water is found in these places. The average air pressure on Mars is actually slightly *less* than the triple-point pressure of water – which favors the solid or vapor state of water and not the liquid state – and consequently for Mars to not be a *water planet*. In fact, no liquid water has ever been found on Mars, even at the polar caps, where there is known to be much water but that is hard as a rock frozen solid.

With atmospheric pressure at or above the average (and it can vary about 50% from place to place such as at the bottom of the Hellas Basin, which has about twice the global average air pressure) and temps above 32 degrees F, water could – theoretically – exist in a liquid state *but only in a very narrow temperature range* at those exceedingly low air pressures, like 32-50 degrees F according to one estimate (science1. Nasa.gov), which means it would boil at 50 degrees F but such a thing has never been observed. The air pressure is so low on Mars that even in places that are known to exceed the average by about 50%, and where it ostensibly would exceed the triple-point pressure, it's still extremely low – 1% or less that of Earth's air pressure – such that at only a few degrees higher than ice melting – it would boil! The boiling point might

---

1   The famous ice-skating example purports to demonstrate the triple-point of water, but this seems to demonstrate not the triple-point but the more general effect of *pressure* on a phase change, which is still provocative because most are familiar only with the effect of *temperature* on a phase change, as with ice melting.

be considered less a constant than it is on Earth and vary depending on the air pressure which, in turn, is determined by the altitude *of the terrain.*[2] Conversely, at any air pressure less than 6.1 millibars (the triple-point pressure) water would either rapidly boil away or freeze (depending on whether it's above or below 32 degrees F). *But* even where the air pressure is higher the hourly temps fluctuate wildly (including 32-50 degrees F) everywhere on Mars, as previously explained, creating *extremely* little opportunity for liquid or standing water. Note that the freezing temperature of water is practically identical on both planets, and doesn't even *have* a boiling point on Mars except for temps above 32 degrees F and pressures greater than 6.1 millibars, at which point the boiling point temperature will increase going from higher to lower altitude locations.

All this means the location of a colony will need to be at as low an altitude as possible, something equivalent to Earth's Death Valley, where the boiling point temperature will be as high as possible – still only 50 degrees F or so – and liquid water might be maintained with *some* degree of reliability and convenience – in facilities that would require round the clock heating but at least that might eliminate one problem, in a somewhat natural way – without engineering – even though the low air pressure *would still otherwise be fatal* to humans without a spacesuit. And next to the north polar ice cap. Where water ice is exposed for parts of the year. Or so it seems.

The fact that water boils (on Earth) at 212 degrees F has enormous applications, from cooking to scientific research and industry. All the relevant standards and techniques would be gone on Mars, where for the most part water doesn't boil! This helps to explain why everything would need to be reinvented for Mars.

But then again even this "triple-point" is just a high-minded distraction, or false flag, after all it's not as if there had been a consensus that there couldn't be liquid water on Mars because of an outdated presump-

2   Science teachers have asked students for years "what's the boiling point of water?" the answer being 212 degrees F. But if a student on Mars was asked the same question, the answer would be "where?"

tion that it's always below freezing, and everyone already associates temps that are above freezing with liquid water – but the fact that on Mars water *boils* at temps just above its melting point (depending on the air pressure) instead of the more practical 212 should be BAD news for the Mars advocates, rather than the good news they're making it out to be, another nail in the coffin rather than a ray of hope.

> **But that's how propaganda works, even those who make it up succumb to it. Under the circumstances, thinking that water might exist in a liquid state on Mars sufficient to support a colony is like thinking that bird hunting would be easier if birds flew at 1,000 miles an hour. So the odds of man living on Mars are even WORSE than anyone would have thought!**

Certainly it is through a grand design – a cosmic conspiracy – and not mental hangups that man could never live on Mars.

The triple-point! Of all things! Who would have thought? The extreme criticality of this can't be overstated! This may be the strongest reason of them all to turn away from Mars, so you can see why it's the best kept secret of NASA and the Mars advocates.

> **Because of the "triple-point" conditions on Mars it's IMPOSSIBLE for water to exist in a liquid state, no matter how much water there might be on Mars as a gas or a solid! Except of course for fleeting moments that might occur, perhaps every day, in various locations which, even if those locations covered vast areas, only for fleeting moments could one even HOPE for water to exist as a liquid!**

So this exception is much more technical than practical – an abstraction. If there were ever liquid water observed on Mars today it would be *phenomenological*. This reminds me of the story of two cursed lovers, where one was an eagle and the other a wolf, but only for half the day, so while one was a wolf the other was human, then they'd convert, so the one that was human became an eagle and the other, human – but for a brief second they were both human – another one of those unrequited love stories. This would describe our relationship with water on Mars, but based on the evidence even this will not happen.

Despite forty or more years of detailed satellite imaging of the

Martian surface, which would surely have detected anything big enough to supply a small colony with a convenient water supply, as would be indicated by the highly reflective surfaces of standing liquid or frozen water – and continuing lack of other evidence – triple-point conditions that exist on Mars continue to tantalize scientists with the possibility of liquid water. But if conditions truly favored, or there was direct evidence of, any kind of liquid water, that would be a major breakthrough and put an end to overly excited mass media news of such things as an "ancient dry lake bed" that occurred in December 2013, and all the other dreary reports. So if today's headline reads "NASA finds ancient lake bed"... that means there's no liquid water on Mars!

Whatever water that exists on Mars is locked up as ice and it would be gone forever if we tampered with the balance that now exists, through evaporation. Deliberately melting whatever ice that exists and exposing it to the atmosphere presents the immediate problem of rapid evaporation. The water wouldn't just evaporate as it does on Earth, it would sublimate, go directly from a solid to a gaseous state, be literally *sucked* into the dry Martian atmosphere. And whatever Martian "water cycle" that may exist would not favor the return of water once it evaporated.

In this light, probing for pockets of frozen water that may be interspersed throughout the surface layer of Mars seems silly, because

Even if Mars was hollow and filled with water, it would be what oil is to us now – and at some point it would be used up, even if it's carefully recycled. Then what? And running out of oil here would be NOTHING like the inevitable running out of water on Mars.

But Mars is not hollow and filled with water.

Alternatively, if Mars was COVERED with water, like Earth, what difference would it make, because there still exists the other very non-trivial deal breakers: an atmosphere that has virtually no oxygen, low air pressure, deadly radiation, no magnetosphere to shield us from the radiation, the extreme cold, the weak gravity – and no natural food supply, all of which the Mars advocates and scientists dismiss as if they were merely inconvenient but each would require insanely ridiculous engineering schemes and *human adaptations* that would never end. So,

Mars has already put us on notice – sure, there may be water, but it's *not liquid.*

On Earth, man has the prerogative, and convenience, to live just about anywhere he pleases, without regard for temperature and certainly not air pressure.

> **Now imagine if our entire population was confined to small fractions of a planet where we sat around and HOPED for the triple-point to occur – and where other resources ranged from restrictive to nonexistent and you needed to wear a spacesuit on top of that! Would you choose that – or Earth?**

OXYGEN

This same exact analogy of using a dollar bill to describe the Martian air pressure can conveniently be used to describe the relative difference between the amounts of oxygen on Earth and Mars: If a dollar bill represents the percentage of oxygen in Earth's atmosphere, then just over half a penny, or six-tenths of a penny, represents the percentage of oxygen in Mars atmosphere (21% of the Earth's atmosphere is oxygen, 0.13% for Mars). You can see the differences between Earth and Mars are adding up, this last one being a doozy.

You like to breathe, don't you?

According to some descriptions there is "no oxygen" in Mar's atmosphere, but let's go with the 0.13% to placate the Mars advocates. Oxygen is the most abundant element in the Earth, as well as in the human body – what does this tell you about our relationship with Earth? Remember what I said previously, that there may be a fundamental problem with creatures from one planet living on another, insofar as its constituent parts are – more than we might ever know –a function of the planet that gave rise to it?

Earth's atmosphere (at the surface) is not only 21% oxygen – this is the ratio of oxygen that we need to survive! Not more, not less. Human physiology is not flexible about this (give or take a few fractions of a percentage point) – less than 21% and we immediately become mentally and physically impaired, which can lead to unconsciousness and, if not corrected, death. Sedentary activities or strenuous, it doesn't matter. But for intense physical activities, such as would be occurring on Mars in setting up a colony – and in their daily exercise routines – maintaining ample supplies of oxygen would involve a special, ongoing sense of urgency.

Oxygen can be mixed with certain other gases, like argon, so the ratio includes about 21% oxygen, but nothing can substitute for oxygen, and it must be 21%.

MARTIAN AIR IS A TOXIC SOUP

Poison gas may be a better analogy, but combining the air pressure and oxygen situation...let's just say that there is no air on Mars. No air on Mars! But let's belabor the issue, since there is that tiny bit. What the Mars advocates would call Martian air consists of 95% carbon dioxide, 3% nitrogen, 1.6% argon, and traces of other gases including oxygen totaling less than 0.4%. Yes, that's zero-point-four-percent. Now let's look carefully at carbon dioxide. The carbon dioxide in the Martian atmosphere is an entirely separate problem from the virtual lack of oxygen. We exhale carbon dioxide – our bodies don't want it, it's a waste product. We breathe in air, our lungs extract the oxygen, then we breathe out the carbon dioxide as waste. Now, our solid waste – our feces – is flushed down the toilet into the sewer, and elements of the sewer system are treated with much disgust and horror – that is something about which we all stand together. By analogy, the air that we breathe out is, in so many words, "sewer air" – and might as well be treated with the same disgust and horror as sewer water! And the Martian air is full of it! 95% full of it!

The Martian atmosphere is almost entirely sewer air! This is just part of the bitterly cold reality that the Mars advocate space cadets militantly ignore and dismiss.

Mars hates oxygen! – the very thing that our bodies crave! You just can't trust those postcard views of Mars – even if they might look like an adorable Arizona sunset.

You like to breathe, don't you?

You will never see these statements on the Wikipedia page for "Colonization of Mars" – or other mass media promoting fantasy and propaganda as scientific fact.

TOXICITY OF CARBON DIOXIDE

But it's worse than just a respiratory waste product. We all already know that carbon monoxide is a deadly poison (it spews out of your car's tailpipe), and guess what? So is carbon dioxide. The Martian atmosphere is 95% carbon dioxide, compared to 0.039% in Earth's atmosphere, or 390 parts per million (ppm) – as one might hope considering its toxicity to us! (This was the percentage in 2012 and has been increasing about 2 ppm per year in recent decades due to human activity.) Scientists say "parts per million" when they're measuring something that exists in very tiny amounts. The effects of CO_2 poisoning begin to occur at about 1,000 ppm, or 0.1% of the air being breathed – fortunately this is just something we don't need to think about on Earth because our levels have never been anywhere near 1,000 ppm – so imagine the 950,000 ppm on Mars. At amounts of just 600 ppm, the air is stuffy and slightly difficult to breathe. At just 1,000 ppm, symptoms of CO_2 poisoning include headaches, hyperventilation, fatigue, and hearing loss. At CO_2 levels of between 15,000 to 30,000 parts per million, severe symptoms include nausea, shaking, hallucinations, vomiting, and loss of consciousness, leading to death if oxygen is not made available. Carbon dioxide poisoning can cause long-term nerve damage or cardiovascular conditions such as hypercapnia. When the blood carries a surplus of CO_2, as with long term exposure to elevated levels of carbon dioxide, a medical condition known as acidosis occurs, where the acidity of the blood increases, blood pH drops below 7.35, and permanent cell damage occurs. So it's not as if carbon dioxide can be treated as a neutral, incidental gas in the mixture of gases that make up the air we breathe, or that as long as there's a certain percentage of oxygen then the amount of CO_2 doesn't matter. CO_2 is toxic in amounts measured in parts per million and can kill.

Plants, of course, need CO_2, but even for plants, CO_2 much above 1,500 ppm is toxic, that's 0.15%, or about 4 times the amount normally found on Earth – and, again, the Martian atmospheric CO_2 is a whopping 950,000 ppm. At toxic levels of CO_2, plant tissue death occurs. So if you thought that a CO_2 rich atmosphere would be great for plants – or that one simply needed to warm the Martian air to make it more favorable to plant growth – you'd be completely wrong.

As far as the gaseous components of Martian air are concerned, Martian air is completely toxic to both plants AND animals.

As you can see, artificial environments would have to be created and maintained for both plants and animals to live on Mars, and at significant pressure differentials. However, this might be impossible in the short, medium, or long term, at any scale.

GIVEN THESE THREE THINGS ALONE, REGARDING TEMPERATURE, air pressure, and oxygen, you'd think this would be a revelation, a major scientific breakthrough! A turning point. A media event. A paradigm shift. On the unhurried and sometimes arduous road of progress, there are stopping points. We are at a stopping point. The headline is written, but not printed – it is suppressed. There are no little green men...there is nothing. Yet the Mars advocates persevere...witless, undaunted.

"There is nothing in the desert, and no man needs nothing."

Despite NASA's own evidence and given their plans for additional future rovers, they must still be expecting one of the rovers to unexpectedly bump into the mythical Shangri-La – absent that there seems no reason they wouldn't just announce the obvious conclusion – which would be *better* than a Shangri-La, quite frankly (imagine the absurdity of Mars tourism) – and finally put an end to this chapter. Otherwise they're just perpetuating the "little green men" myth, if in a different form – but such "scientific objectivity" is too much to expect even from our brightest.

There is, of course, another way to look at it.

There's more money to be made in not knowing than in knowing. Ambiguity keeps the money flowing to colleges and industry in an effort to more precisely measure...the same old facts. But which is all so very unscientific – and unethical, since there is more value, ultimately, in embracing the truth than in promoting lies like "Mars is habitable."

But there's more.

GRAVITY

The gravity of Mars is one-third the gravity of Earth. The human body reacts pretty badly to space travel and low gravity, with both muscle and bone deterioration, in some ways that are irreversible. This does not pose the same imminent risk of instant death as with the issues concerning oxygen or water, but presents problems that are just as fundamental to our well-being concerning the short and long-term prospects on Mars. An artificial habitat and spacesuits might protect man from the other deadly circumstances, but it is not possible to compensate for the reduced gravity – without great effort and large machinery but it seems any strategy that involves this would make for a miserable existence because a life can't be lived in defiance of nature. Which means if we're on Mars, we'll have to live with its reduced gravity.

It is a big unknown whether Martian reduced gravity can support human life in the long term, and the long term consequences of living in Mars' reduced gravity can't be predicted. According to one theory of space medicine researchers, sleeping chambers built inside centrifuges would minimize some health problems. The Mars Gravity Biosatellite experiment was due to become the first experiment testing the effects of partial gravity, artificially generated at 0.38 g to match Mars' gravity, using mice, throughout the life cycle from conception to death, but in 2009 was cancelled by NASA due to *lack of funds*. This experiment was intended to see if mice could reproduce in microgravity, which would indicate man's prospects to do the same – and which should be of great interest to NASA but for some reason IS NOT! NASA changed their priorities – to what? Is this a subtle hint that they really aren't interested in a colony on Mars? Or are they avoiding the facts – hiding from reality?

RADIATION

UV radiation – which is extremely harmful – readily passes through the thin Martian atmosphere, which is less than 1% of Earth's, and with Mars having an Earth-like schedule of days and nights is impossible to avoid. This means Mars has extreme amounts of solar radiation and in combination with everything else renders the surface of Mars inhospitable to life as we know it.

Radiation of the ultraviolet kind will pose a greater hazard on Mars than the threat of a global nuclear war on Earth.

One could say the unfiltered UV radiation on Mars, in quantities far greater than found on Earth due also to the relative lack of a magnetosphere and which would pose an ongoing and serious threat to anyone on Mars, is warfare of another kind, but just as deadly and mutating.

MARS DIED 3 BILLION YEARS AGO

Expert opinions vary, but the consensus is that Mars died 3 billion years ago (which does not necessarily preclude continued geologic activity like volcanoes or tectonic plate activity). Its core froze, its volcanoes seized up, and solar winds stripped away its atmosphere. All the liquid water evaporated or soaked into the ground and froze, leaving the surface too cold and dry to support life.

There are, evidently, vague indications of methane gas vents on Mars (with trace amounts of methane on the order of 1 or 2 parts per billion – which is less than previously thought and considered to be "no methane" – detected by the rovers [Science, 9/19/13]). This is another one of those things scientists get real excited about, because there is life on Earth, and methane on Earth, so we associate the two, and think that this same association must exist wherever we find methane even if we find methane and no other signs of life and other signs of life are completely contraindicated! When it comes to Mars, scientists are always making these leaps, which is misguided because methane also has geological sources, which are far more likely with Mars than biological sources and which have nothing to do with life. Any compound on Earth could be associated with life on Earth because we know there is life on Earth! You could therefore say that any compounds on Earth also found on Mars could be a sign of life – but wouldn't this seem hysterical? Methane is not a universal sign of life just because it is associated with life on Earth.

THOSE ARE THE FACTS. DO YOU STILL THINK MARS IS HABITABLE? None of these facts will change in one, ten thousand, or ten million years. The laws of nature of a planet, whether that planet is Earth, Mars, or any other – for all practical purposes – don't change. It seems *reasonable* to conclude that Mars would not be a nice place to live – indeed, that Mars is not habitable in the least. And it's unreasonable for us to continue believing that an environment that is not conducive to human life would somehow contain the answers to human life. A few might someday be able to visit, but visiting a place and living there are two different things – a *world* of difference since we're talking about another planet. A lot of people have a "we can do anything" mentality, even though we can't, of course, do just anything. Especially if that means that by ignoring the facts they will go away. Or that the belief that humans are "tough and resilient" means that we are literally made of steel! Or that being omnivores – which we are generally considered to be – means we can literally eat anything and with no regard to flavor or nutritional content – because this certainly isn't the case.

> **But all these things would have to be true for us to live on Mars. We'd have to be made of steel, and omnivores in the literal sense – but obviously these can never be.**

So, how do some conclude that Mars is not only livable, but "hospitable," as some are claiming – which means warm and inviting – at least that's what the dictionary says but maybe there's a Martian dictionary that says hospitable means brutal and toxic, or something along those lines, which is what Mars is. I've got to get my hands on one of those Martian dictionaries, then maybe I'll understand how anyone can think that Mars is "hospitable." Otherwise, if we agree that hospitable means "warm and inviting" then there's a whole lot of propaganda and brainwashing going on and I'm not quite sure why.

There's certainly a truth in advertising issue created when selling this to would-be space travelers, if you're selling them a life of adventure and riches – as is the case. I can't imagine anyone who would even consider making such a trade off.

Astronauts to this day haven't had to make such a trade off, and joyfully return to their earthly existence.

The trip to Mars would be one way; once anyone acclimated themselves to Mars, they could not likely return. Their physiologies would have been altered that much, perhaps even their anatomy, especially those born on Mars. Our life-span would probably be affected, our life-cycle. Development would be completely different than on Earth.

Even if there were some planet in our own solar system that was an exact replica of Earth, there would be SOMETHING that prevented us from living there, something only discovered after a few manned missions, but a proverbial deal breaker nonetheless – the potential culprits would be endless, ranging from the mundane to the exotic. We already know Mars offers far less than such a fictional Earth replica.

The table on the next page summarizes this chapter, and includes other information future space tourists will need to know *in advance*.

Every single vital parameter describing Mars is a worst case scenario or creates a serious and ongoing difficulty for living on Mars – call it a living Hell – no, Hell is better!

In my worst nightmares, I could never imagine Hell being this bad! There are many situations or options FAR better than this for which I would prefer death. This Hellish nature of Mars can't be overstated.

Any one of these parameters by itself could be considered to make Mars uninhabitable – certainly all of them combined. But the Mars advocates call this *hospitable*. Of course they're mentally ill! – and I make a lengthy case for this in a later chapter. Even if three or four of these parameters were Earth-like, it wouldn't matter, because the others remain. Even if ALL of these parameters were Earth-like except only one, the overall effect would still be "unearthly."

Only someone who is mentally ill could look at all this and call it "habitable."

Feature	Status
Water	Worst case scenario – no liquid water anywhere, water ice at poles, but won't help if mid-latitude temps most conducive to a colony.
Oxygen	Worst case scenario – things couldn't be worse – deadly.
Carbon dioxide	Worst case scenario – things couldn't be worse – deadly.
Food	Worst case scenario – things couldn't be worse – nothing on planet that can be used as a food source.
Radiation	Worst case scenario – 100% avoidance of direct exposure to sun at all times – must wear spacesuit at all times.
Temperatures	One notch above a worst case scenario, with almost constant subfreezing, occasional temps above freezing depending on latitude and time of day.
Gravity	One-third that of Earth, causes muscle and bone loss, which can be partly compensated through exercise.
Natural Resources	Worst case scenario • No oil or materials that can be used for fuel or to supply energy (or impossible to contrive on the spot) – except for solar energy (which could be used to charge batteries but which would need to be brought from Earth – another one of those impossibly difficult inconveniences – bearing in mind that battery performance is hindered by cold temps). • No liquid water. • No animals or plants for food or byproducts. • No trees, therefore no wood, therefore no furniture or structures or other things made from wood. • Since no plants can grow in freezing ground or weather (besides the other hostile environmental factors) there will be no cotton and therefore no cotton clothing. • No sources for other natural or synthetic fibers for clothing or other things.
Distance	Worst case scenario – Would require rocket technology that doesn't currently exist and which may never exist, but even then rockets could only be launched once every 26 months at enormous expense and risk.

Even good old Earth is a source of constant surprise, and can be quite the antagonist, what with earthquakes, hurricanes, monsoons, tsunami, and epidemics. Collect all the data and take all the pictures you want, the true scale of Mars is beyond the grasp of remote cameras and sensors, where strange and hostile phenomenon lie in wait.

But let's ignore all the science for the moment – it's inconvenient for those of us with passion and a zest for exploration. We want to live on Mars and nothing is going to stop us. Right?

4

The Spacesuit

The spacesuit is a Procrustean Bed.

BUT THE SPACESUIT WILL OVERCOME ALL THIS! WHAT MORE could a man want? Give me a spacesuit and...life is grand. Whenever I see man depicted in some alien, sci-fi landscape – an artistic rendering – a warm sense of reassurance comes over me when I see that he is in his spacesuit. His trusty spacesuit which, by itself...gives man superpowers! This is how simple-minded the Mars advocates are, and is perhaps the greatest assumption of them all, after the rocket. Give me a rocket and a spacesuit and we can colonize Mars, no problem – the Mars advocate mentality can be reduced, almost, to this degree of cuteness.

The spacesuit is a given in every vision of man in space or on Mars, but it's not simply taken for granted. The spacesuit is a *Procrustean bed.* [In Greek mythology, Procrustes, in order to fit travelers into his iron bed, would cut off their legs or otherwise stretch them.] They're unbelievably complicated, incredibly difficult to get into – several people are required to get into one – and obviously extremely uncomfortable, their bulkiness imposing huge restrictions on not just general mobility and dexterity but severely handicapping all five senses – like a modern day Procrustean bed. First, they just might exceed the degree of technology that would be maintainable on Mars in the long term. They're inherently too high-tech for the low-tech Mars man may one day be living on – imagine cave men wearing spacesuits. Well, that does it. Oh come on, not so fast. OK – but they're not designed to be worn for long periods of time – I mean helmet and all, hermitically sealed. Being in one is not much different from walking on the bottom of the sea in an old fashioned diving suit, the one with the big metal helmet that's screwed on from the outside that predates the modern SCUBA outfit and look how clumsy *they* are. To think that anyone could or should "live" in a spacesuit is outrageous – no matter how modernized it may be.

A spacesuit handicaps the senses? Regular clothing handicaps the senses! What other species wears clothing? Other than to stay warm during cold weather, no article of clothing enhances the human experience, and people almost always, other than for ceremonial purposes, wear the minimal amount of clothing necessary either to stay warm or to stay *legal*. There is certainly nothing equivalent to the relative monstrosity of a spacesuit that is in your or my wardrobe, and this is as true for any previous generation as the present.

The irony is that the original lunar spacesuits, as worn by Mr. Armstrong, weren't even meant for physical activity! – and provided almost no range of motion! So to think that the spacesuit is not an issue is downright foolish. It is as big an unsolved problem as any of the others.

The spacesuit would also interfere with perspiration. Evaporation is a cooling process and sweat evaporating off your body cools you down. This is why we perspire and for no other reason and with an airtight spacesuit this physiological process is completely defeated. If you're in an airtight spacesuit, where's the sweat going to go? It certainly isn't going to evaporate and cool your body down, and you wouldn't be able to just open a vent in the helmet or in the suit to compensate for a faulty cooling system (because that would cause instant decompression of the suit, and death) so whatever heating and cooling systems are built into the spacesuit will need to be fine-tuned and failsafe. This will obviously be much more a concern on Mars than for our earlier trips to the Moon because unlike our brief visits to the latter, lasting from only hours to a few days, man would be living on Mars in a much fuller sense and engaged in labor intensive activities as he sets about building a permanent colony. And since the temps can easily swing 100 degrees or more on a daily basis (plus 30 to minus 70 F as a random example) being comfortable in a spacesuit will be a matter of life and death. People on the ISS don't wear spacesuits, since it's pressurized, so this issue with normal perspiration isn't a problem – or as big a problem – on the ISS.

If man was starting from scratch on a new planet, one might hope for an ideal in which clothing of *any* kind was...unnecessary! Either as protection from the elements or out of the contrivance we call *modesty*. But a spacesuit? That needs to be worn 24/7 – and without which one would immediately die? This alone should define Mars as *uninhabitable*!

But like all other practical considerations, let's put that aside.

With a pressure suit, very short-term scouting around may be doable, assuming suits that are comfortable and allow freedom of movement, and impervious to wear and tear since even a slight rip or puncture could cause almost immediate death – no one is designing suits of this nature at this time, but it's an enormous assumption to think humans could live out their lives wearing such an outfit even part of the time – it would be an insult to the normal physiology and kinesiology/ergonomics of the *body in motion*. Any pressure suit would need to be imperceptible to the wearer, a *second skin*, to make the whole thing doable, and this needs to be first generation, not tenth generation, and let's face it – it would need to be sexy.

Even when you weren't actually wearing it, you'd have to handle it like a Fabergé egg because just in the process of taking it off or putting it on tiny cuts or tears might appear in the material, maybe to one of the interior layers of fabric – but which wouldn't be noticed. This would be more likely to occur to those involved in industrial or manufacturing activities such as digging up Martian dirt or making or laying bricks, or any other work activity – or play or leisure activity – it wouldn't matter. Or falling or being knocked down onto the Martian ground, which would scrape against the suit, such as with Martian youngsters who might be playing or horsing around outside, bullying, etc. Any kind of cut or tear, no matter how small – even invisible to the naked eye – or any other flaw or defect in the workmanship of the suit...would immediately expose the person to the deadly Martian environment. After donning the spacesuit and stepping outside, they might notice something is wrong right away and have time to step back into the shelter. Otherwise, if they didn't notice the drop in pressure or other problem right away and had walked some distance from the shelter and couldn't make it back in time – or hadn't even realized there was a problem – they would probably first collapse, become unconscious, and then die, all within just a few minutes. If someone else wasn't right there to provide immediate help they wouldn't have a prayer. And to the naked eye the suit might look just fine, even brand new. The entire suit and helmet would have to be discarded because it might be impossible to locate whatever tiny rip or tear or unnoticeable manufacturing defect was the culprit, even if that was suspected to be the case *and it probably wouldn't be.*

Is it possible to make a spacesuit that is impervious to even the tiniest pinholes or rips or tears and that can withstand the rigors of both work and play, while providing freedom of motion and uncompromised use of all of our senses? The suit worn by Neil Armstrong sure didn't accomplish this – and is the main reason he was on the Moon for only two hours! And you probably couldn't substitute for the loss of sensation by building it into the suit – that would make it far too complicated and even more prone to malfunction.

So the integrity of the spacesuit has far ranging implications regarding the safety of the individual colonists and the colony as a whole. But other things need to be taken into consideration also: features of the suit that would identify the wearer or otherwise intended to convey authority or rank (which might be as simple as an armband), design aspects related to the gender and age of the wearer, aesthetic aspects of the suit design – all these may be incorporated into the suit, or attached in some other way. Otherwise all the colonists will look alike (out in the field all wearing their suits) and they couldn't readily identify the doctor or the mechanic or just each other. For example, if you have a group of workers lifting something heavy into place – which would probably be done manually rather than through heavy machinery, starting out – the foreman needs to know who each of the workers is in order to call out instructions – which would be impossible if they all looked alike in generic suits. The individual identities of the workers would need to be obvious, to stand out. Among other things, helmets should have a life-size picture of the owner's face on the back so they can be readily identified from behind.

The helmet interior would need an array of tiny but high quality speakers positioned around the head so if someone is saying "Help me with this" from 30 feet behind you, that's what it sounds like, rather than turning left and right in confusion to figure out who might be in trouble – but at the same time you don't want to hear every conversation going on among dozens of workers at a work site. The communication system built into each helmet would need to allow large groups of colonists, say, at a work site or gathered closely to inspect something to have private conversations with each other. Perhaps in beginning a sentence with the person's name, only that person could hear, or with "private one" only the one person closest to the speaker, and other simple code words to control who hears what. Such complexity wasn't a concern with "Man on the Moon" for obvious reasons, where a simple walkie-talkie design was sufficient. So each suit would need sensors to detect the distance and direction of everyone else out in the field. But as with other cool ideas, this may be too high-tech for the low-tech Mars that would be inherently less prone to malfunction. Ideally, the sound system in the helmet would need to provide a realism that includes environmental noises like the sound of an approaching dust storm, and other things that we are as yet unaware of – or to selectively filter out such noises. All of which begs the question: always wearing an air-tight helmet, will it be possible for anybody to actually hear the sounds of Mars?

These and a million other considerations have to be made – just with the trusty spacesuit – for even a colony of just a few dozen, working in close cooperation, to succeed on Mars.

SOME MARTIANS WILL INEVITABLY BE CARELESS, RECKLESS, OR have a death wish – but that might be true for anyone who wants to go to Mars! Anyway, humans are notoriously careless about taking care of themselves. We overeat, drink, smoke, abuse drugs...do we think these things wouldn't happen on Mars? Look at all the motorcyclists who don't wear helmets, despite the obvious risk. Martians, too, will just get tired of wearing a helmet, take some other shortcut, go outside and that's it! I'm just stepping out for a second...whoops! Sometimes it does get above freezing on Mars. On such a "nice" day, someone is bound to take their helmet off, just for a few seconds – it's bound to happen. They may even have contests to see who can stand up the longest without a helmet, and other dare-devil *adventurous* ideas. People will just get tired of all the dressing up and dressing down. For large populations, it's expecting too much to think that each and every one will be "properly outfitted" at all times, with helmets, gloves, booties, straps, zippers, buckles, air canisters (which alone will involve a litany of rules and regulations) etc., and everything properly double-checked. You wouldn't even be able to go out for a walk by yourself – people would be required to travel in *threes* and being with just one other person would be *illegal*, the danger of the Martian atmosphere is that extreme.

Bear in mind that the spacesuit as we know it was designed for Neil Armstrong to basically stand like a statue on the Moon for *two hours*. So this or anything like it would not do for man living 24/7/365 on Mars. The spacesuit would play a much bigger part in living on Mars than it did for the lunar astronauts – or than it does for those on the International Space Station, who don't even wear spacesuits while on the ISS (except for the brief moments one or two of the six might step outside to do some critical maintenance such as to repair a leak).

The results of a contest, sponsored by NASA, were announced May 1, 2014 (on the ABC Evening News) for the design of a spacesuit to be worn on Mars – on which was emblazoned an enormous, glow-in-the-dark "Y." It doesn't get more Freudian – I couldn't have said it better myself! *Why*? But of course. If this whole thing was a card game between me and the Mars advocates it would seem they're giving me the best cards on purpose.

So, the spacesuit may not represent as big a challenge as the breakthrough needed to land a big, cargo carrying ship gently on Mars or the whole issue of getting back to Earth...or does it?

5

The Mars Advocates

There's a religiosity to the Mars advocate mindset,
reminiscent of the suicidal Heaven's Gate cult that
sought immortality via the Hale-Bopp comet, or the
UFO they believed trailed behind it.

MAN HAS, SINCE THE MIDDLE AGES AT LEAST, BEEN INTRIGUED BY
the idea of life on other planets, and sometimes we – that is, science-fiction junkies – imagine what it would be like to live on other planets.
During these same centuries, right up to the present, many have been
preoccupied by thoughts of travelling to these distant planets. Mars is
closer than any of the other planets, so if conditions are like those found
on Earth, some think it might be a feasible alternative to living on Earth
– have even made the leap that conditions on Mars are "hospitable" –
even though it seems remarkably clear that the opposite is true! Let's
call these people the Mars advocates.

Let's also agree upon a few simple definitions – at least, I would
expect them to be simple. In the most basic sense, a place is "habitable"
if generations of people can spend their entire lives there without being
preoccupied with simply surviving. This obviously does not describe
Mars. Many places *on Earth* do not satisfy these conditions and are
therefore considered to be uninhabitable.

"Hospitable" is, of course, a special refinement of this, such
that issues regarding survival disappear altogether and
where one's sense of well being becomes transcendent –
which certainly does not describe Mars!

And what does "survival of the fittest" mean if not the contest between living things and the environment that's trying to kill them? In
which case it would be a stretch even to say that *Earth* is hospitable.
But Mars? If Earth were truly hospitable we wouldn't need houses.
The magnitude of the investment we make in our houses (and all the
other machinery of our lives) is testament to and a measure of the effort
of our struggle to win this contest – in which case it would seem that
Earth – Earth! – is inhospitable to humans *in particular*, since we're the
only species that needs to be completely sheltered to survive *because it's
certainly not for the sake of being fashionable.* But...Mars?

"For years we've had our eye on the skis." This sentiment dates back
to the time of Galileo and before, when people thought the Earth was
the center of the universe, and later on the sun. It sounds harmless but
it has become the moral imperative of the Mars advocates, who cling to
such weather beaten sentiments as an indisputable rationale to not just
visit, but to actually colonize Mars! But the dreams and myths of our
ancestors would obviously be reshaped as the centuries rolled by – one
would think, since the advent of science, its influence, its irresistible
omnipresence – so this wake-up call will be a rude one:

> **Wake up, Mars advocates! Time for the paradigm shift to
> begin. We've learned a lot since our cave days, including
> the fact – the very indisputable fact – that Mars is a dead
> and deadly planet. Once upon a time we thought there
> might be little green men on Mars, but now we know
> there aren't. Going from "I have a rocket" to "Let's colonize
> Mars" isn't a moral imperative, it's certifiable insanity.**

One of the cornerstones of the Mars advocate logic is the belief that
Mars is more like Earth than the other six planets and we should therefore move there. This is analogous to a poor person standing in a yacht
store convinced he should buy the cheapest yacht because the others are
too expensive. The cheapest yacht would be the "least worst" option to
a poor person, likewise Mars is the "least worst" of all the other planets
to a Mars advocate. Affordability per se is not the point. Rather, in
the case of the Mars advocate, optimism, hubris, and an unexplainable
desperation convolute a "least worst" choice into something desirable,
even vital. Both are examples of the same epistemological fallacy. In
relative terms, the cheapest yacht is, literally, more affordable, but a

poor person could never buy a yacht, and for obvious reasons. Even if a poor person won or inherited a yacht, he could never afford the expenses of using or maintaining it, including fuel expenses, berthing fees, etc. (If you would argue this example, you just might be a Mars advocate.) It's not a coincidence that poor people don't own, or even want, yachts; evidently, hubris comes into play only as one advances on the scale of wealth, as with the more entrepreneurial Mars advocates. Because it is pure hubris for a man to think he can live on Mars – who doesn't just ignore all the reasons that prevent it, but thinks he can defeat the laws of nature, including human nature, that prevent it.

> **The Mars advocates, along with NASA, might be described as being Panglossian, from the character Doctor Pangloss in Voltaire's "Candide" – characterized by or given to extreme optimism, especially in the face of unrelieved hardship or adversity – the unrelieved hardship or adversity being the unlivable conditions on Mars.**

It may be well within the bounds of human nature to seek to discover and explore other planets, but to conclude that Mars is livable, knowing what we know now, gives new, unearthly meaning to *Panglossian*.

This reminds me of a poem by Arthur Golden.

> *Adversity is like a strong wind,*
> *It tears away from us all but the things that can not be torn,*
> *So that we see ourselves as we really are.*

What might cure the more entrepreneurial Mars advocates, with their degrees and corporate pedigrees and the hubris that goes along with it, is a little adversity in their lives, that is, a bit of normalcy. So they might better see themselves as they really are. But of course that wouldn't be enough.

Logical next step?

The suggestion that Mars is the "logical next step" seems more pathological than logical. It seems that man would do anything to AVOID colonizing Mars – which seems like the most extreme scorched Earth solution conceivable.

The Mars advocates think it's easier to move to a planet that has no oxygen! – than to change our bad habits (the nuclear threat, overpopulation, etc.). Rather than to simply stop making nuclear bombs, which would be an obvious way to address the nuclear threat, it would be easier to redefine the laws of nature of another planet, according to Mars advocate logic – reading between the lines at first but which becomes clear from the body of literature that continues to accumulate, in the form of sci-fi novels, magazine articles, and internet websites.

Anyone who wants to move to Mars should already be living in a freezing desert, which would be a simple thing to arrange – Antarctica or Siberia – but even this would be exorbitant compared to what's waiting now for them on Mars. Such a life would also be paradoxical – insane – in a world obsessed with upward mobility and hospitable climates.

It seems one reasonable "logical next step," which should address at least one of the Mars advocates' concerns – regarding overpopulation – would be to implement any form of population control one would care to imagine, no matter how outrageous, because that would still be more rational than colonizing Mars. It would make more sense to colonize any of the deserts on Earth, which would not require the extra step of terraforming. Or does the idea of colonizing Mars presume this will have happened – that even our deserts will be overpopulated at the point we "take Mars." If the fear is that we've run out of space here on Earth then it would make more sense to tear down some existing cities and rebuild them! Of course this sounds ridiculous but no more than the idea of colonizing Mars. Many cities suffer from years of misguided urban design. Tear them down and rebuild them as sleek, modern cities, with a Martian flair. Let the Mars that many imagine...be here on Earth.

It would make enormously more sense to develop uninhabited areas of the world, or America – places that are all relatively *hospitable* compared to Mars – New Mexico or Arizona, Death Valley, the middle of the Sahara – places that provide unseen opportunities for the growth of new cities or other development, rather than our stubborn tendency to expand existing cities – places that have oxygen! All in the spirit, of course, of a new world order, featuring genetic and biological "adaptations" to suit the Mars advocates' fancy, drawing from and fully exploiting all of modern man's resources and talent pool and free of the usual bureaucratic or "slacker" influences. Work on this for a few years...which turn into decades...and they will "inevitably" have, I think, that indescribable thing they're looking for – or at least a version of it that has oxygen and water – and quench that burning, pent up, insatiable *je ne sais quoi*. The Mars advocates truly need to rethink the idea of "habitable," as a place that can support a flourishing human society on the order of many thousands of years, before taking another step toward Mars. And for those who would take that trip, be warned: there is no Holiday Inn on Mars. There are – besides no oxygen or water – no conveniences, no creature comforts, no amenities. Just inconvenience, discomfort, and a punishing reality that is unlike anything you could possibly imagine or dream.

Here's one way to put it: if there were already a Holiday Inn on Mars with all the amenities, a hiking trail to Mount Sharp, and an underground souvenir shop, everything else on Mars being the same, with a round-trip excursion to the red planet where each way takes 8-9 months out of your life and all you could do on board the rocket is read and watch movies (with a drug-induced coma option) – all free – would you? I can't imagine anyone wanting to do even that, much less if the trip was one way with no coma option and nothing but brutal and toxic desert landscape waiting for you. A week in the middle of the Mohave without water is more inviting, and certainly more accessible – at least the weather is nice and you could play with, and eat, the scorpions. Here on Earth try looking for a "desert vacation package." I mean into the pure desert. That might indicate future Martian prospects, but there are none (other than the oasis in the desert kind, where you're really not in the desert).

And another way: imagine a town by a reservoir, with a dam holding back the waters, and that dam had a hole in it, and someone stood there with their finger in the hole to save the town from certain destruction. And everyone in the town took turns standing there with their finger in the hole. In fact the role of everyone in the town was to take turns doing this. Would you want to live in that town? How desperate would you be to move to that town? Would you consider life in that town to be an adventure? Rather, how desperate would you be to AVOID living in that town – LIKE THE PLAGUE! But that's what it would be like living on Mars! Do anything possible to AVOID living on Mars!

OBSESSIVE COMPULSIVE DISORDER

The Mars advocate mentality might be described, by the mental health industry, as an OCD (Obsessive Compulsive Disorder) rather than just another fringe group. Or perhaps they're giving expression to an atavistic need, a throwback to our medieval ancestors, stupefied by the sparkling night sky but with little choice in the matter because they didn't know any better, couldn't know any better. Certainly this atavistic preoccupation is, in light of what we know now, a curse and not a blessing. Granted space can be exciting and enthralling, but at some point shouldn't the well reasoned mind prevail? Or do we just allow ourselves to be hustled with little regard for cost, benefit, or danger?

There's a certain religiosity to the Mars advocate mindset, because it obviously represents much more than jobs – or even the next incremental step for (ahem) science. It's so very reminiscent of the Heaven's Gate cult that started out in the 1970's with the two founders touring America and sharing their New Age beliefs about the afterlife and *parallel universes.* They believed that passage to the afterlife could be gained via a UFO they believed trailed behind the Hale-Bopp comet, which came closest to Earth in March 1997. (Among a group of 39 followers who committed mass suicide at this time in order to meet with the UFO was a brother of the actress who played Uhura in the original Star Trek TV series. [Wikipedia.]) But have the Mars advocates surpassed this, such that their zeal can be compared to that of the Reformation during the Middle Ages, or to that of Mideast based terrorism? NASA, of course, as well as the Dutch Mars One, gives the whole thing an authoritarian edge.

True Mars fanatics such as Robert Zubrin, founder of The Mars Society, wax quite prolific in a somewhat insane rubric about rocket science, politics, cost, sustainability, and even profit from commercial ventures, where no problem can't be solved with "engineering," while completely excluding from consideration the biological, emotional, psychological, social, and ethical issues. To The Mars Society, it's a foregone conclusion that people exist for no reason other than to be *workers*; that people are nothing but DRONES serving no purpose other than LABOR. You will find in their literature nary a mention of our personal interests, hobbies, recreation, or leisure activities, and no mention whatsoever of family life, all of which would foretell an existence on Mars that is relatively sociopathic – depraved – by Earth standards. And so it seems that life on Mars, even through the voices of its most enthusiastic promoters, would be a sterile, joyless existence.

IF IT'S A CASE OF SEEING THE UNIVERSE THOROUGH ROSE-TINTED glasses, and the belief that there is something better out there – or even a barely livable substitute – there isn't – at least not within our reach, despite the illusion of closeness that evidently occurs for anything seen through a telescope, in combination with our natural capacity to wonder and imagine, which in turn is inspired by countless science-fiction fantasies, which just might be the holy trinity of modern astronomy but which results in a complete blurring of fact and fantasy where some believe that warp drive is *just around the corner*.

> **However, there is good news. We can make this world a better place – we just have to separate the illusions from reality, embrace the reality, stop chasing windmills, and be pioneers in the one place in all the universe that chance gave us the luck to be.**
>
> **Creation has imprinted us to be Earthlings. We are geniuses of Earth, and at the same time its masterpiece. That is the grand design.**

MENTAL ILLNESS?

There are those who will say I need to be more open-minded or optimistic. This is all part of the kook mentality, which has no place in space exploration. The kook mentality predates the Middle Ages and has wound its way through the centuries to modern times. It manages to survive in one form or another and has completely taken over modern society – it has become the dominant voice! NASA – which has had a preeminent role in bringing us into the space age – also, unfortunately, provides a bullhorn for this voice, as does SpaceX, Mars One, and celebrity astronomers. But science is neither pessimistic nor optimistic, and has no feelings. It's when feelings enter the arena that things get kooky.

Knowing what we know now about Mars (no oxygen, water, or food), the following argument can be made: The belief that man should travel to other planets is a weird thing. It must somehow be compensation for something, in those who have this belief, but especially where this belief is a preoccupation or obsession as it is with the Mars advocates. Everybody has their "thing." It might be alcohol, or sports, or sex, or food, a career, kids, material goods, a pet, vanity, cars, money – all of which are somewhat normal, and seem to satisfy most people's sense of fulfillment. But (incredibly) some people don't have any of these things – they resort to fulfillment in more unusual ways which by definition suggests a maladjustment, or even a maladaptation. This is not necessarily a bad thing, where a profit or some other benefit can be realized. But there's an even more extreme direction, taken by some, that seems entirely pointless and misguided – a *severe* maladjustment – by which I am of course referring to the Mars advocates. This is not simply a manifestation of a leisure society that has nothing else to do.

BUT PERHAPS I'M NOT GOING FAR ENOUGH. PEOPLE WHO THINK travelling to other planets is a good idea or who want to go to Mars specifically must surely be mentally ill, right? And I don't mean because they lack any comprehension whatsoever of the loathsome cost/benefit ratios that are involved. *Knowing what we know now*, about Mars, the continued belief that humans can live on Mars...is deranged. You can't celebrate the ideology because some part of it is capitalistic, and creates a few jobs – because on the whole, it's deranged.

> **Let's at least agree that the whole thing has become passé, and that it's time to move on in a more provident direction – but this may take a revolution.**

At the one end, many people are gullible and prone to suggestion and brainwashing – let's call them space cadets. As we move along the spectrum, where we find many Mars advocates, there are those who think they can fly (who are either stoned or mentally ill) or are preoccupied with some other "superpower," in which case they may be retarded or otherwise mentally handicapped. Further along, with increasing dysfunctionality, there are those who have "other worldly orientations or ideations" and who need to be medicated. Therapies and other countermeasures already exist for similar mental illnesses or are being investigated. Many categories of mental illness are actually far less onerous or debilitating. And so it seems due time that this mental affliction, or complex – which may be considered to define, at least in part, the Mars advocate – be addressed by the mental health community. It's really not such an unreasonable proposal – the number of people affected, and severity of symptoms, certainly seems to be clinically significant. We can start with the 72,000 people who already signed up with the Dutch Mars One starting April 2013 (although only 7,000 included the application fee) and the numbers that followed (from which a few will, according to Mars One, be chosen for Martian astronaut training) – who have unwittingly outed themselves! We've got their names. The mental health community needs to put the proper diagnosis on this, such that "Mars advocate" be considered a bona fide psychiatric condition. *Knowing what we know now*, of course, about Mars.

EVEN BEFORE WE SENT THE FIRST UNMANNED CRAFT TO MARS — long before even the first of four rovers – we knew what the gravity was, what the oxygen levels were, the temps, the day and night cycles – in other words, whether or not we could live there – and the answer was... No. There are four rovers on Mars, "roving around" since 1995 and sending back data (the first two stopped transmitting in 1998 and 2010). There has been found to be a COMPLETE absence of life on Mars – how can the Mars advocates not be impressed by this? What are they thinking, that some as yet undiscovered fossil of the only bacteria to ever live on Mars will somehow represent a green light for Man's next giant leap? When will we get the "all clear" that it's OK to move there – Rover LXII? The rovers have at least seemed to confirm that there are no Martians – which is strangely disappointing. So it seems we should finally give up on the Mars thing and turn our attention elsewhere: to Earth. Imagine if all the ingenuity of getting the last rover to Mars could have been focused on solving unemployment, world hunger, the collapsing global economy, global warming, or any number of other incurable issues. One needs to question our most general priorities if we've given ourselves over to *mental illness*.

To a man of modest means, ambition can be a curse, but nowhere is this more evident than in the Mars advocates.

LET'S TAKE AN EVEN CLOSER LOOK INTO THE MIND OF THE MARS advocate. What motivates them? For one, the Mars advocates fear that, in one form or another, Earth is on an inexorable path to destruction, and that mankind will be "saved" by moving to Mars. Various apocalyptic scenarios are imagined, both natural and man-made, including species extinction due to asteroid impact, mega-volcano eruption, ice age, and others – as if we would be totally free of all of these on Mars. Mars has its own natural disasters, including planet-wide dust storms and vulnerability to asteroid impact (including a big one predicted in the vicinity of Mars about 2014 that, if it hits Mars head on could have apocalyptic consequences for both Mars AND Earth).

One of the more ridiculous delusions of the Mars advocates is the belief that they will be safe on Mars while the rest of us live in a state of peril on a doomed Earth. No matter how you look at the facts, no matter how one might skew the facts, no matter how bad things get on Earth,

the survival of humankind – aside from any guarantee of comfort or convenience – will always be easier on Earth than on Mars. Because the fact is, no future Earth apocalypse is as bad as Mars is now!

If you want to envision a dead, burned out Earth – just think of Mars! Earth will always at least have oxygen, water, vegetation, etc. (radiated or not) whereas

on Mars we'd be starting from absolute zero! Actually worse – since, by analogy, Earth minus humans and the grand total effects of humans would be "starting from zero" and Mars is obviously much worse off than that.

This is perhaps the ultimate fallacy of the Mars advocate ideology. Even with a so-called extinction level asteroid hit, like the one that may have "wiped out" the dinosaurs – which isn't actually true in consideration of alligators and other "prehistoric" species that appear to be *thriving* – humans have far greater – more numerous and strategic – life support mechanisms than the poor dinosaurs.

> **Some humans would probably survive even worst case scenarios, if only as mutants – but is this worse than the mutants we would certainly become on Mars? Better to be a mutant on Earth than a mutant on Mars – the devil you know and all that. But it seems all this might be resolved by...wearing spacesuits now!**

A Martian colony would not, obviously, be "life as we know it" – let that be perfectly clear – or somehow better. And of course the colonization of Mars wouldn't actually save you and me personally.

If a nuclear holocaust is one of the Mars advocates' fears, it seems that "uninventing" the nuclear bomb would make more sense than colonizing Mars – an implausible idea to be sure, but implausible ideas are the specialty of the Mars advocate – they can use their limitless imagination to figure out a way to uninvent the nuclear bomb, which is really no more farfetched an idea than terraforming Mars.

The nuclear bomb threat is a prime example of our self-destructiveness. Would this human quality just disappear on Mars? Nuclear technology already exists on Mars – it's there waiting for us. The thing we fear the most on Earth – one thing which we theoretically could "undo" to start anew – is already waiting for us on Mars! We put it there! Some of the machinery on Mars, onboard the NASA Rover Curiosity, is nuclear powered. Some will claim the particular isotopes involved are not "weaponized" but it's a small leap from that to nuclear bomb. Some will argue that the isotopes used to power the Mars rovers can't be weaponized but that's hardly reassuring. Marie Curie didn't die in Nagasaki. But aside from man-made isotopes,

> **radiation of the ultraviolet kind will pose a greater hazard on Mars than the threat of a global nuclear war on Earth.**

One could say the unfiltered UV radiation on Mars, in quantities far greater than found on Earth due to the relative lack of a magnetosphere and which would pose an ongoing and serious threat to anyone on Mars, is *warfare of another kind*, but just as deadly or mutating. So if the Mars advocates think they would be escaping a nuclear threat, there's nothing to be gained on Mars.

Some of this could, of course, be explained through "mortality salience," in which efforts to colonize Mars would be a form of "terror management" and thereby provide greater meaning to life. This would have nothing to do with one's sense of adventure but could go a long way in rationalizing the Mars advocates' desperate psychology and gullibility.

IF SELECTIVE BREEDING IS SOMETHING WE MAY NEED TO DO ON Mars, why not selectively breed ourselves RIGHT NOW to get rid of our self-destructiveness? Because that would be unethical. But why would it not also be unethical on Mars? Because selective breeding would be absolutely necessary on Mars, to hasten the process of adaptation. But isn't selective breeding absolutely necessary here on Earth for equally fundamental reasons? And so we could go in circles, the point being we could make this planet, Earth, a better place to live – Now! – were it not for our self-defeating and self-destructive – yet inextricable – tendencies. Just why do we think we can somehow escape these things – or anything else – by "moving to Mars?"

Humans are too profound to be moved around like a potted plant.

Humans are not something simple, like a potted plant that can be moved from one window to another. We are not just another element in a simple equation. The human element...is really 10,000 elements, an unsolvable matrix of variables that are both known and unknown. We are not one, but a collection of phenomenon of indescribable complexity. We are borne of, and tied to, our environment in ways we still don't completely understand and might never. Even within its tiniest subcomponent, human life remains profound – too profound to be dispossessed – evicted from – its home, like the proverbial fish out of water, to think it can continue to survive, as if whatever it is that binds us here – to Earth – are petty hangups or something ridiculous, as if our wholeness is separate from it. In moving to Mars man would experience a total loss, a kind of death. One can't experience a "total loss" and act as if nothing's different – as if our milieu is created by us for our convenience. Quite the opposite: it would seem undeniable that we're a *manifestation* of our environment, and one would think this would be understood, if only slightly, by everyone. But *the engineering mentality in particular, in some cases, seems to lack this intuition* – doesn't simply not understand it, or deny it...but LIVES to deny it and deliberately, conscientiously, insists that it does not exist!

> **It's interesting how the culturally stunted engineering mentality, where it exists, takes this route – to use technology – rather than the route that requires skills of human interaction and compassion such as for people to get together and strategically decide, through patience and negotiation, to live in harmony.**

These engineers always think their latest invention is the sine qua non or deus ex machina, but their inventions are toys and gimmicks and never really change the basic nature of life, and end up instead comprising the landfills that surround us. Moving to Mars is their latest gimmick.

Even if there were to be *any* degree of success with a colony on Mars, one can still insist that the Mars advocates are mentally ill. Some will find a way to force a square peg through a round hole and then call it ingenious, and it may always be a matter of opinion whether any size colony on Mars is successful or not. *How would anyone on Earth ever really know what was going on up there anyway?* The Mars advocates will claim success regardless, but in the end it will be impossible for anyone to prove that their beloved adventure – and whatever other magnificent payoff – had ever been realized.

6

Elon Musk - Part 1

The justification for "colonizing Mars" is not something that can be based on a cult of space cadets, nor high riding entrepreneurialism or capitalism, not the dream of a few rich megalomaniacs or PT Barnum clones in the form of Elon Musk and other rich and crazy space junkies hawking space travel, or even the fake altruism of "creating jobs." Rather, it is a fundamental truism that whether or not man sets out to colonize Mars should be based on hard-edged sanity, not delusional hopes and dreams – or psychos.

IT'S EASY ENOUGH TO IMAGINE GOING TO MARS. WELL, ACTUALLY, it isn't, but let's pretend – for the sake of humanity. There are many other things I'd rather imagine but, continuing, our spaceship lands on Mars, and we check into the local Holiday Inn, where we have reservations (imagine along with me). Wouldn't that be great? Well, not really, but TRY to imagine, or suspend disbelief, or whatever else you need to do. But even while imagining we need to be careful because this isn't like going to the Moon. We never intended to live on or colonize the Moon or to mine its resources, although popular science magazine articles had everyone believing that – which now seems indefensible. Colonize the Moon – how ridiculous! Authors of pop science articles still write with the same reckless abandon, regarding many things (predicting the demise of books, etc.) but on the subject of outer space, writers have switched their focus from the Moon, to Mars! Crossing man-made boundaries into foreign or unknown territory may be bold and daring, but to escape the ultimate boundary of them all – this is indeed provocative. Going to the Moon was a joy ride, but Mars is a completely different story.

OK, stop imagining. The idea, and the consequences, of moving to Mars seems even to elude the grasp of scientists, as if all one needs is passion and a validated boarding pass. It's not as if we're going to get off a spaceship and walk to our reserved room. You'd need a lot more than just a base camp to accommodate people who are moving there for good – which seems to be the plan, according to Mr. Elon Musk, founder of SpaceX, the rocket building company in California. Perhaps Mr. Musk should quit while he's ahead, having already revived serious interest in electric cars in the form of the more or less successful Tesla model he helped fund and which first appeared on streets in 2008 – even Michelangelo had only one Sistine Chapel. Mr. Musk is, based on popular press and respectable science articles alike, a megalomaniacal opportunist (or successful venture capitalist, depending upon your choice of phrasing) who seems to have decided *it's time.*

Elon Musk is a successful entrepreneur who developed SpaceX in 2002 to design rockets to service the International Space Station (ISS) as it orbits Earth with its international crew, and which his company has succeeded in doing, using the Falcon and Dragon rockets which he and his team personally designed. In this way SpaceX assumed the role of the Space Shuttle – and just in the nick of time because that program was retired in 2011 despite its success over 30 years (or not, if you consider it's 40% vehicular failure rate). But Mr. Musk has a greater vision – he wants to save the world – by colonizing Mars.

It's incredible that someone starting out as a web designer – PayPal – could reinvent himself as a rocket designer – but it's true. That's quite a career, and his biography will be a best seller. He is the proverbial "one in a million," and if he stops now, he will have fulfilled a potential most of us dare not even dream.

But the truth is, SpaceX is nothing but a more cost-conscious counterpart, or extension, of the Space Shuttle program – and which could be considered, in every other way, a downgrade, since the Shuttle program clearly had a lot more potential.

Let's be clear about one thing: SpaceX is not, nor its Dragon and Falcon rockets, an evolution – especially not technically – and Mr. Musk himself has said as much, in numerous articles and press releases. We're no closer to Mars – not even a baby step...no closer to saving humanity. Mr. Musk is no genius savior, even if he has managed to live off the exploits of that crazy internet capitalism which has served him well – in a big way with all his weird luck! Because it's the same old rocket science! SpaceX is the same old rocket science but with a lot of big talk and geek appeal. It's a different group of engineers, and a different zeitgeist – guided more by the spirit of "lean and mean" than the old school rocket science, whose resources were more unlimited...*but this only addresses the economies of scale that crippled the Shuttle program, not the evolution of space exploration on the grand scale.*

Mr. Musk doesn't know what he doesn't know, and clearly does not care about the details of man living on Mars – according to his public statements – but what can you expect from someone described – in the popular press – as *weird*? (Smithsonian magazine, December 2012). He apparently hasn't assembled a diverse team of experts, nor sought to reduce the mountain of what he doesn't know. It's just him and his *rocket scientists.*

MARS IS AN AIRLESS, SUBFREEZING DESERT PLANET. A READING of the science-fiction novel *Dune* may be in order for anyone who subscribes to the Martian Earth substitute fantasy. If imagining yourself living on the fictional desert planet Arrakis, portrayed in *Dune*, doesn't provide some insight, then please continue reading. At this point in time, Mr. Musk and his company, SpaceX, are designing a vehicle, and trying to sell tickets to ride that vehicle, to a destination that is largely imagined but that doesn't really exist and most likely never will.

Thinking we can live on Mars if only we had a rocket to get there is like thinking I can get rich in Vegas if only I had a car to get there. (I have a car, of course – I'm just making an analogy.)

Of course there would be a certain thrill in flying a rocket to Mars – for the first few hours anyway – which is about the length of a weekend car trip – at which time you'll start hearing the first refrains of "I want to go home!" Have you ever wondered why the roller coaster is so thrilling? A ride only lasts 2 minutes! You don't really want to live on a roller coaster for eight months! – which is the time it might take to get to Mars in a SpaceX rocket.

It's ironic that somebody who lives near Malibu – coveted for its climate and scenery – final destination for the hip and the hot – would even think of such a strange thing as...moving to Mars – as if this was part of some weird, new age progression: Malibu, then Mars. Or maybe it's only too appropriate that this idea comes from a place sometimes known as La La Land.

And of course JPL, premiere authority of the space industry and command center for the Mars rovers, is right there, along with SpaceX, in the same neighborhood as Hollywood, entertainment capital of the world. The space industry, other than those parts of it responsible for launching or maintaining satellites and other useful things, is arguably just another facet of the Hollywood/Southern California entertainment/ dream industry. To this extent SpaceX, NASA and its offshoots and subsidiaries are in the entertainment business, if not in the usual sense. One lab is actually working on chemicals that would enhance the flavor of prepackaged food bound for space – from which perhaps no one may ever benefit. Maybe they could consider the hungry in Africa, who number in the millions – or isn't that another kind of saving humanity?

Have you ever seen a football player after losing a game, during the post-game interview, look into the camera and say, flat out, that "the other team is better?" No player could ever concede to this fact, or that it's just a game over which they have absolutely no control whatsoever. Instead, they beat theirselves up, analyze it, over analyze it, make promises, lay blame. This same hubris is in abundance at SpaceX, NASA, and JPL – you'll never see any of them look into the camera and say "Sorry folks, game over. Mars is not suitable for human life." There's just too much money and careers at stake. The dream machine is just too profitable. And perhaps it's accurate to say that this particular dream has proven intoxicating. Which is to say nothing of the effort and time being spent by all the other engineers, and engineering students, and collateral industries working on projects focused on the space industry in general or living on Mars in particular, or technologies being cannibalized for this purpose. One can only wonder if we've past the rubicon already.

How can SpaceX *not* be affiliated with Disney, in its mission to indulge the thrill-seeking public by creating what amounts to nothing more than yet another thrill ride which, granted, would be an enormous leap forward in terms of craziest diversion ever. Feel the G-forces! Ride the rocket to Mars!

The proposal, or implication, that the mission of SpaceX is, directly or indirectly, to "save humanity" is difficult to buy in the presence of this other, immediate gratification, thrill-seeking mentality.

One of the hardest things for people to realize, for those who apply for Mars astronaut training, is that Mars has one thing on its mind: to kill anyone who would presume to live there.

When celebrities like soprano Sarah Brightman claim to be interested in going into space or Mars (and who has an album with outer-spacey themes titled *Dreamchaser* which, ipso facto, seems to make her a Mars advocate – she may in fact epitomize the persona), they are nonetheless being used as tools by Mr. Elon Musk or other space promoters, to raise their respective profiles and increase their celebrity status. Because you can be sure that Ms. Brightman knows nothing about space or Mars and isn't about to step foot outside of her chauffeur driven existence.

As she states in a June 2013 interview, regarding her "singing on the ISS with an orchestra," "it all depends on the practicality of the technical aspects."

Bravo – I couldn't have said it better myself. Sarah Brightman is, in fact, the perfect tool for the Mars advocate propaganda machine. Propaganda often works by saying things are the opposite of what they truly are, so the bright and sunny Brightman plays perfectly into the deception that Mars, too, is bright and sunny. Her name alone says "bright man." Are you in shock? I personally would like to be the first to perform *Yakety Sax* in space – that would scare away the aliens I'm sure.

7

Elon Musk – Part 2

*Mr. Musk's infinite means and infinite ambition
spell infinite hubris, which does not equate to a
colony on Mars and which could quite possibly
lead to infinite disaster.*

IN NOVEMBER A.D. 2012, MR. ELON MUSK OF SPACE-X FAME MADE
an outrageous proposal to an attentive gathering of the Royal Aeronautical Society, asking them for money and support in what may be considered one of the greatest frauds of all time: to put a colony of 80,000
people on Mars. Mr. Musk claimed "the project would be designed to
make the settlement increasingly self-sufficient over time." Therein lies
the principal epistemological fallacy of his plan, and is in plain view: dependencies on earthly resources couldn't be sustained in the meantime,
before self-sufficiency is achieved – because of the enormous distance,
of course. It is axiomatic that a colony would need to be self-sustainable
from the start – as soon as the first crew landed. Even at first glance, the
proposition Mr. Musk made to the Royal Aeronautical Society, which included the claim that anyone who wants can get to Mars, is both bizarre
and ridiculous *in the extreme.*

Getting to Mars isn't the problem. The rocket science isn't the problem. Getting rockets to land gently on Mars *isn't the problem* – even
though this is impossible now and a significant technical aspect of the
rocket science – but *not* one of the more fundamental problems with

living on Mars. As if anyone would go to Mars, stay for a few months, then return to Earth but for the inability to *land gently*! **It's living on Mars that's the problem!** Mars is uninhabitable, with no oxygen, liquid water, or food, among other deadly deal breakers described in much technical detail elsewhere in this book.

One aspect of the dream logic of modern space promoters like Elon Musk and Robert Zubrin, leaders of the Mars advocate mob, may be the most intriguing of all. The expertly curated individuals chosen to make the trip (let's say 100 but however many seems impossible to predict) will pay for the privilege...the figures start at $100,000. That's right, PAY! And, according to Zubrin in *The Case for Mars*, our adventurous space travelers will be only too willing to do so! Zubrin has worked out the details. There will be various financing options available, of course, not unlike for a college degree or home mortgage. But will they be making money on Mars – like with a regular job? Now don't whip out your currency converter just yet, because of course there will be no compartments stashed with cash on the rocket ships back to Earth – but once returned, the astronauts will experience "fame and celebrity"... which translates into money, naturally. I can barely imagine 100 astro-nauts turned celebrity, much less 80,000. I'm sure they can celebrate each other as they throw back calcium supplements to undo the bone loss...poor things – *things* – will be lucky if they can resume any kind of natural life. And will they be bringing back any baby Martian progeny human-y things with them?

But it's utterly outrageous to recruit 80,000 people to live on Mars!

If there were already a colony on Mars, growing slowly but steadily over 40-50 years to a size of, say, a thousand – somehow defeating the odds – and then a miracle that allowed people to live comfortably – equivalent to how people live on Earth – at that point a case might be made for some long term urban planning, to include a city of perhaps 80,000 people. But zero to 80,000 – on present day Mars – can't be defended. The usual description of a colony as a platform for scientific research seems more like code for "seeing if man can live on Mars," which would be an obvious prerequisite for a larger colony, with the question remaining of what the colonists would actually be doing, besides merely surviving.

To realize a colony of 80,000, you would, at some point, need rockets that could carry at least 100 people, but such a rocket would be enormous and involve unacceptably high risk factors. Alternatives to this would include staging such events from the Moon, where certain efficiencies might be realized since the gravity is much less. But this would require a full-scale lunar colony to house the astronauts, and a way to shuttle astronauts from Earth to the Moon – which would cancel out the efficiencies to be realized from a lunar launch to Mars, plus the costs involved in this would be enormous and issues of practicality would at some point intervene – where science-fiction becomes *bizarre fantasy.*

But getting to Mars isn't the problem – nor efficiencies regarding this! It's living on Mars that's the problem!

Mr. Musk doesn't really care about colonizing Mars, or the suitability of Mars as a home for mankind – at least not in the typical Mars advocate sense. Clearly he has no foresight into what would be involved for humans, or anything, to live on Mars. According to his own statements, he has no interest in the midrange or long term future of humans on Mars. But it would require many rockets to transport 80,000 people to Mars – *his* proposal – so just think of the potential profit for his rocket factory, SpaceX – and him personally – if such a vision were to be realized – a potential destined to be exploited by the entrepreneurial Mr. Musk despite the foolishness involved and there just may be no end to that. But in promoting this idea Mr. Musk becomes not only a Mars advocate, but a voice for the Mars advocate propaganda.

Actually, the idea of "just flying a rocket to Mars" isn't that outrageous. It's practically ordinary. NASA – and space agencies in other world powers like Russia – has been successfully launching rockets to Mars for decades, since 1960. *But they're all unmanned.* So it's not outrageous that Mr. Musk would simply want to launch his rocket at Mars. I used to launch (model) rockets at my local school yard, it was fun – so launching a really big rocket should be even more fun!

But it's utterly outrageous to recruit
80,000 people to live on Mars!

Especially if they're paying as much as half a million bucks each. People who have that kind of money enjoy the luxuries of Earth way too much to say good buy to them, and people with that kind of money gravitate to a lifestyle (impatient, being waited upon, power, status, fancy cars, fancy furnishings, fancy food, luxury vacations, fancy attitude) that doesn't customarily lend itself to being an astronaut – or *becoming a Martian*. People who have that kind of money have mastered the laws of Earth-nature and the rules of the game of capitalism; they've made a 100 percent investment in their existence on Earth and demonstrate an economic survival of the fittest, in which lies their success, their payoff – and which also inextricably binds them to Earth. For anyone willing to go to Mars, Mars would be a kind of prize not understandable to the rest of us, especially not the high earners Mr. Musk seems to expect to recruit. Because life on Mars would consist of (among other horrible things) digging up Martian dirt to make bricks and other very labor intensive activities that would fill their days.

Mr. Musk must think there's a nobility in going to Mars, or that he's fulfilling some grand destiny. It's a vision of some kind. He's serious...or seriously gone mad. It's one thing to build a rocket...and launch it. But move to Mars, not just man, but Mankind? Is Mr. Musk a visionary...or a huckster, a modern day P.T. Barnum...or an antichrist? To get 80,000 on Mars, many will obviously die in the process, but this is all part of the calculus. But this "many must die" mentality, or calculated risk – call it what you want – seems unchristian, even anti-Christian. There was supposedly a greater good, according to some, that was to come from the Holocaust too – meaning it's all a matter of perspective, isn't it?

ANTICHRIST

Could anything good come from this? Is SpaceX's Elon Musk some form of antichrist or psychopath? He certainly seems to be leading us to some kind of Hell. The Mars Mission is the Holocaust redux – one can only too easily create an analogy between the Holocaust gas chambers and the toxic carbon dioxide "gas chamber" atmosphere of Mars: equally deadly, equally maniacal. Jobs creation? Human size gas chambers created jobs but that wasn't a good excuse for the Nazis, but if you're a high flying entrepreneur it – and the equivalent *Martian "gas chamber" atmosphere* – becomes not just a good excuse but a matter of national pride? A form of genocide? – one way to get rid of the most fanatic Mars advocates, that's for sure! Reverse genocide, where you kill the ones everyone thinks you're trying to save – diabolical.

Mr. Musk must think a human being is just another element in a simple equation, a simple machine that can be simply put in its Procrustean bed. [In Greek mythology, Procrustes, in order to fit travelers into his iron bed, would cut off their legs or otherwise stretch them.]

Mr. Musk doesn't understand that the human element... is really 10,000 elements, an unsolvable matrix of variables that are both known and unknown.

The problem with Mr. Musk is that he has a rocket and doesn't know what to do with it! It's befitting that Mr. Musk is a rocket scientist, because of all the engineering specialties, rocket science is the one most irrelevant to the human condition – so this would at least partly explain his mindset. One might also say rocket science is the least necessary of the engineering disciplines, and not particularly vital to my or your well being, certainly not in any immediate sense, unlike with the relatively vital engineering disciplines relevant to electricity, bridges, highways, energy, chemicals, materials, biomedical equipment, etc. Whereas these engineering specialties have a focus on not just the quality of our daily lives, but also the progress of mankind and in a very direct way – the same can't be said for rocket science, certainly not as it pertains to the bizarre goal of *travelling to Mars*. No aspect of life on Earth requires rocket science, and this includes all technologies made possible by satellites, which we've only had for the last 50 or so years and which are frivolous from the standpoint of our species survival.

The use of rockets in warfare alone, and in its perpetual threat to peace, may cancel out all the benefits.

It would make more sense if Mr. Musk just built factories to make electric cars for everyone – since he does know something about electric cars – and gave them away – as a less ridiculous public works project. A free electric car for everyone – FREE! That would make the world a better place in more ways than one and may actually be about 100 times less crazy than sending 80,000 to Mars.

Certainly Mr. Musk has greatly overestimated the prospects for a colony on Mars, and also the number of people who might actually go. Among those who work at NASA and the space industry – the most obvious recruits – most have no interest whatsoever in actually going to Mars – if any – *even if they say they do*. And if I'm wrong – that everyone who worked for NASA and the space industry fully intended to go to Mars – there'd be no one left at ground control!

8

Training 80,000

As if a colony on Mars somehow starts the countdown
for the rest of us left behind on Earth to perish in some
inevitable "end of days."

THE BIGGER QUESTION MAY BE, HOW DO YOU TRAIN 80,000 PEOPLE
at once – based on the outrageous proposal of Mr. Elon Musk of SpaceX.
And this would only be the "first wave." Try to imagine the facilities that
would be required. Consider the scale and complexity of the facilities required to train just the few dozen lunar and space shuttle astronauts over
the entirety of those programs. Even assuming efficient use of facilities,
so that they were in operation 24 hours a day, with training facilities in
each state, each state would need the facilities to train about 1,600 people
– unless of course you wanted to "go global" but maybe a Made in the USA
approach would be best since that would better serve our national pride.
The 1,600 people being trained in each state would need to be distributed among at least 10 facilities per state, each facility hosting about 160.
Bigger states would of course have more training facilities than smaller
states, but even the facilities training fewer people would need to provide
the same quality of training, such that you'd have at least one "G-force
simulator," or whatever, per training facility – or 500 nationwide (10
per state for each of the 50 states). But there are only a few of these in
existence now! I'm sure they cost a few million a piece – $50 million?
$100? And you'd need several per facility to train several at a time and
minimize downtime. Then you would need 500 "deep water simulation

pools," 500....of everything. There are barely people alive that would be required to support such an enterprise! The factories that would need to be built, the machinery that would need to be maintained... In a facility equipped to train 160 there would be a training and support staff of at least that many, which equates to about 80,000 training staff nationwide, probably more. And of course you'd have to train the trainers – none of whom have ever been to Mars (at least initially). The entire enterprise would dwarf the automotive industry. It would of course be a huge expense, subsidized by federal and state governments, *and all while not providing any real goods or services*, other than to provide jobs (which might convince some "nonbelievers" to switch sides).

There would, of course, be no regard for expense, this being one of those "public works" projects – for the greater benefit of "the citizens" and all that, in the true Keynesian spirit...or might this more likely be given over to something more totalitarian? There are many variations of this architecture – make one up!

But anyway, it would take years just to construct this enterprise, whose sole purpose would be training people to go to Mars. Can 80,000 be ready for the "first wave?"

This is, of course, a flight of fancy, since such an enterprise would never and could never be built – or maybe it is in fact an entirely intelligent plan!

But rather than attempt this colossal build-out we might build it into existing schools and colleges, and make it available to all students... which could potentially provide many Mars recruits, over time.

But do we really want to give ourselves over, en masse, to such a totalitarian ambition? This would make it seem that the entire point in being human was to become Martian!

Even if we discovered that the seeds of our existence on Earth somehow originated on Mars – then got here via a long and cold ride through the vacuum of space on a chunk of debris from some cosmic collision – it seems the investment Earth has made in us since then has precluded the possibility that a successor should come easy – and vice versa: the investment we've made in being earthlings.

It doesn't make sense to think that our society would back a Mars mission in some totalitarian sense – which is what it would take – because since 99.999% of us would always be on Earth, our individual survival unaffected by a colony on Mars, the only thing that would make sense is that our society would back a "Project Earth" in some totalitarian sense! In an Earth vs Mars future, it seems obvious that Earth would get the majority vote! But my instincts to survive preclude such decision making, even in the theoretical sense – as if I need more choices! Only in a movie theatre would I be capable of the "suspension of disbelief" that the Mars advocates are asking of me. As if a colony on Mars somehow starts the countdown for the rest of us left behind on Earth to perish in some inevitable "end of days."

9

Recruiting

Of course no one who is making money – receiving funding – in this arena will be going to Mars, be the proverbial rat in the maze. Of this you can be sure!

THERE SEEMS TO BE A REMARKABLE LACK OF INFORMATION RE-garding what the first colonists on Mars would be doing, and each successive wave of colonists, and therefore a general confusion regarding the evolution of a colony. But this is important to know because the nature of each successive stage of a developing colony would be quite different from the previous, with equally different roles for the participants, running from scientist to laborer to all manner of technical specialist – with a heavy emphasis on the labor side starting out. Since they're already recruiting it seems this information would be not only available, but widespread – in a way that would include not just niche websites run by the Mars advocates but also primetime TV and radio ads repeated throughout the day, such that the idea of going to Mars would become ordinary and expand the recruiting base. Without this, how can anyone know whether they'd be interested or not? This can only result in a randomness to the whole recruiting process – so it seems things are already getting off to a bad start.

No part of your daily life – none – is designed or intended for the "long term benefit of mankind," as if to spite the principle of natural selection. Rather, our daily lives are guided much more by immediate gratification and short term benefits. Any advice to the contrary would lie somewhere between intrusive and abhorrent. In other words, if there were a government policy that restricted the size of your family, or the kind of car you drove, or your choice of career, or where you choose to live...there would be a revolution. We would call that a dictatorship – which is anathema to all Americans or those who embrace Freedom and Liberty, and who fully enjoy the principle of Free Will. With that in mind, it is simply implausible to think that people getting on a spaceship bound for some distant planet, perhaps even in another star system, could be told *"to have many children but you'll need to sacrifice a few in the name of natural selection and some of the others will be unusual"* without there being an uproar. Because if that's such a necessary and good plan – which would have to be the case if the Mars advocates have their way – why don't we use the same logic to manage our lives now? Why do the Mars advocates, who promote space exploration and colonizing Mars in particular, implicitly believe that we will think differently when we're "in space" – that our ethical standards and beliefs will be completely thrown out the window? For one thing, it's preposterous to think that any spaceship would have the resources to care for unhealthy or defective individuals – knowing our high maintenance requirements – and including those born on a spaceship – it completely defeats the point of even the most relaxed astronaut screening process.

There's a certain paradox that would be involved in becoming a Martian colonist. It seems that those most interested in going to Mars would include rocket scientists and people in the space and related industries. But it also seems that the knowledge and motivation that propels anyone toward this vision, as embodied by this same group – would be useless once they realized it! Is it possible that the last thing that should be on Mars, trying to start a colony – is a bunch of rocket scientists? Because once they got to Mars, they would be "regular residents," not astronauts trained to fly a rocket. Once there, they'd become farmers and construction workers – laborers – which are the furthest things from rocket science!

An amazing paradox! – since it can be argued that NASA, as much as any other corporate culture, fully embraces a home-body existence, since engineers tend to have the highest paying jobs along with job security – being at the top of the career food chain – which enables them to have nice homes and a reliable, *permanent* address – which would seem to thwart the same sense of adventure that they promote with the other hand. What is a 9-to-5 job and a 30-year mortgage but a repudiation of a life of adventure?

> **Certainly Mars would be the ultimate paradise –
> to a Luddite.**

It's impossible for anyone on Earth to not be disappointed with Mars, especially if they're trained to think it's like Earth, or that it will be a great experience, or some other propaganda. Because while they're training and preparing on Earth they will be psyched and gung-ho and slaphappy, but what's that old proverb about the anticipation exceeding the realization? No matter how much enthusiasm one has starting out, it will eventually wear off. Everyone will inevitably return to a "normal" state – and realize the horrible reality of what they've done. Even if you could implant a microchip to keep that part of the brain excited or intrigued, that part of the brain would eventually burn out.

> **It seems the gung-ho adventure seeker type that would,
> at first glance, make a good recruit wouldn't at the same
> time lend itself to being confined to the close quarters
> they would experience on Mars, nor the "forced labor"
> that they would experience on Mars and that would be
> unending, nor the insulting lack of extracurricular activ-
> ities that, at least for the first wave of settlers, would be
> unending. Once on Mars, all that gung-ho to get to
> Mars would shift into reverse – into an insane desire to
> return to Earth.**

Despite all the eccentrics and weirdoes among us, there's no human archetype that would lend itself to living on Mars. There is no psychiatric condition nor physical ailment that would be cured by living on Mars, as with people with arthritis or asthma moving to Arizona for the dry heat. No one could be so gullible that they would fall for the sales pitch – other than, of course, current or ex-NASA employees who've OD'd on the company mission, who may indeed be overworked and just need some fresh air – or perhaps a bona fide detox. Based on what we've known since even before the Curiosity – the last of the four rovers to land on Mars as the world watched – the continuing belief that Mars isn't just a doable, but a preferred alternative to Earth as a home for humans, is nothing less than a symptom that should interest the psychiatric community. As a pretext to raising nationalistic pride (which is itself a contrivance in a land that seems to consist primarily of disconnected free spirits) it's a nonstarter. It's even more disturbing when one realizes how many of the Mars fanatics – are scientists!

Most professions correspond to particular personality types, for example – and without getting overly complicated – accountants have accountant personalities, pharmacists have pharmacist personalities, and so on. It stands to reason that this would also be true for wannabe Martian astronauts. It might even be the case, and ironic, that many Mars advocates will not be qualified to travel to or colonize Mars. They may actually be the least qualified group of all! So who should recruiters be targeting?

Keep in mind that the most important thing in a recruit is – supposedly – a sense of adventure. Rather than highly educated and credentialed scientists or military, the best recruits may be (hold onto your hats) bricklaying gym rats who eat nothing but canned spaghetti – who else would be agreeable to exercising the two hours a day necessary to minimize the deterioration caused by low gravity, and the spartan menu that would be inevitable on Mars. Try to get anyone else to exercise two hours a day. The dumber gym rats, since they'd also be more agreeable to do the hard labor of excavating the Martian top surface to get brick raw material, making bricks, and laying brick – or farming, preparing the Martian dirt for crops – and not be bored by it or the lack of earthly diversions. The first astronauts, Armstrong and Aldrin, didn't face this predicament because their entire trip to the Moon and back was just a week, only spending two hours on the Moon. One could say they didn't need a diversion on the Moon – the Moon was the diversion, but this will be the opposite on Mars. The early astronauts took no chances when it came to maintaining their earthly bodies. They never intended to stay on the Moon – it was the journey, the glory, and all that rocket science made manifest.

Wait – shouldn't that be bricklaying gym rats who eat nothing but canned spaghetti – and who are also doctors?

Let's pursue this gym rat idea, because the need to work out on a daily basis would in fact become a biological imperative once anyone left Earth (to counteract the muscle and bone deterioration that inevitably occur in reduced gravity – bone loss is even harder to prevent) and would narrow the field of applicants to the Mars program considerably because most people don't like to exercise and are in fact quite sedentary, so this one thing will be *the* deciding factor for many. Imagine if your employer required that you and all your fellow employees had to exercise for two hours a day, 7 days a week. And that if you didn't he would, not fire, but kill you. This analogy is not an exaggeration. Yes, regular exercise will play an enormous part in the life of anyone in space or on Mars – as it does on the International Space Station (ISS). If this doesn't interest you, DO NOT APPLY. And despite the sometimes derogatory connotation of "gym rat," anyone who works out (runs, cycles, lifts weights or some combination of these) on a daily basis is a gym rat! But two hours a day? This would narrow the field considerably.

There are two kinds of people: those who like to exercise, and those who don't. With this dichotomy there is no gray area and once people take sides it tends to be a lifelong commitment. This fact of life must be the starting point for any well focused recruiting program.

But this also creates an obstacle, one might even say a conundrum, and which is hard to appreciate. The priority of gym rats is not working out for its own sake, nor even to be healthy, but body building, which requires a perfect balance between diet, exercise, and rest. And gravity. This is true even for those who emphasize their cardio routines with such things as running and cycling. This perfect balance would obviously not exist in the zero or reduced gravity environments of space, or on Mars, and where the usual benefits of working out would be next to impossible – would not be realized if they were always in a spacesuit, on a spaceship, or confined to a cramped and artificial habitat on Mars – so there would be no point in body building in space or on Mars. Even with simulated gravity, the amount of food available on Mars would not be sufficient to provide the *mega-nutrition* required of the body builder. This means a body builder would need to give up body building in order to go into space or move to Mars, and for most gym rats this would be a DISINCENTIVE. A gym rat would be loath to losing muscle mass – and bone mass! And gym rats tend not to be adventurous, rather hanging out at the gym and sticking around town, preoccupied with diet and nutrition, their 9-to-5 jobs – things not usually associated with being adventurous. In fact it seems that Mars or anything related to it would be the last thing that would interest a gym rat. So what else can one conclude that the only way to convince a gym rat to go to Mars...is by brainwashing and lying to them?

> **Certainly anyone who wants to go to Mars has been brainwashed, lied to, or is just plain stupid. Or they believe they have no prospects for success on Earth and see Mars as an opportunity. And these will be the breeding stock for our descendents on Mars – emissaries of Earth, ambassadors of Humankind.**

THE WRONG STUFF

They would be, one might say, the antithesis of the Right Stuff. The Wrong Stuff. Because it seems that the greater virtue would be to prepare Earth for the future – as if this doesn't require bravery – rather than Mars, which seems to have no potential at all. It's interesting that the *right stuff* (popularized by the book) would be equated with masculinity, since the objective of an astronaut is to leave Earth (where all the woman are) and go to places (like the Moon) where no women are. Do you see the irony? To go where there are no women – or anyone – is both antisocial and asexual. Most would object to this stark objectivity, although in this new context many words have taken on new meaning. The right stuff may have gone to the Moon but it will be something else entirely that goes to Mars: the Wrong Stuff.

If one were to predict the point in time when going to Mars might be the inevitable next step, that might be when there is, for example, a natural population of thousands of dull, bricklaying gym rats – extra dull – living in Antarctica – these should be the most likely to survive – and who have that worker bee mentality apropos for the Martian milieu – and NOT having a sense of adventure. That might at least answer the question of *who*, among the matrix of other unsolvable variables that is Mars. This would also indicate the culture of the first colony on Mars.

That Mr. Musk must be smoking some fine weed to have the visions he's having! Who do you know that doesn't have a spouse, a family, pets, creature comforts, material addictions, a functioning brain – all of which are quite respectable excuses to demure any invitation to Mars. The simple truth is that people have houses and they enjoy being in them. Certainly only a true misfit would qualify to go to Mars, a sociopath – but should this be the *progenitor* of all humans on another planet? Mr. Musk may need to recruit from death row – who else would consider Mars to be, in some form or another, a prize? Or someone who's had the pain and suffering centers of their brain completely disabled. Perhaps there I've stumbled upon it because the perfect candidate can't be rationalized. Surely the Wrong Stuff – right for Mars but wrong for Earth.

Life on Mars would be a veritable prison. Just considering the absolute basics, Bernie Madoff has it better – the convicted ponzi schemer. He's comfortable and in exchange doesn't have to lift a finger; his room and board are paid for, what a life! I just stated that the Mars advocates could recruit from the prison population – but maybe not! Even they would not make such a trade!

SECOND THOUGHTS

The astronauts will, of course, take with them to Mars mementos of their lives on Earth, cherished belongings that might fit into their carry-ons, even contraband smuggled on board. Pictures rolled up in tubes, photo albums, a bag of their favorite candy. But would all these things be nothing but a constant reminder of what they left behind – and would anyone on such a venture *want* to be constantly reminded of what they left behind? How do you leave all these things behind – the lunar astronauts never had to do this. The men who went to the Moon had no intentions of staying there and every intention of returning to Earth, but the Mars mission will be a completely different story. Could memories alone, of their lives on Earth, in some unexpected way, compromise or sabotage the mission? Books, music, and movies brought to Mars may, with all their references to earthly customs and cultures, somehow undo the brainwashing necessary to get them there!

> **Just during the 8-month flight to Mars this will certainly be happening, if not to all but a few, their minds spinning, writhing in horror, as they slowly begin to realize the enormity, the finality, of what they're doing. Because once they're on the way, their days will no longer be filled with the hectic busyness of, not just training, but astronaut training, and all the testosterone and fist-pumping that goes along with that, and that special camaraderie, where they wouldn't have a minute to really think about what they were doing.**

> **But once they're on their way...they'll have nothing to do but think. On Earth, there's really no such thing as "one-way" – you can always go back. But the trip to Mars is ONE-WAY, baby! There will be NO TURNING BACK.**

It will be interesting to see how many astronaut trainees go right up to the point of lift-off – then back out at the last minute. *It will happen.* They could continue their medical training and become a doctor on Earth. Or simply join the ranks of "Project Earth" advocates as they come to their senses.

BORED TO DEATH

What a lot of the Mars advocates don't understand is that doing something isn't the same as imagining it, such that we can imagine what it must be like to do some thing, but there is some aspect of doing that thing that will be different from what was imagined. Because to actually do something requires work, sometimes a lot of work, sometimes risk, or a sense of risk, sometimes even a little blood, sweat, and tears and sometimes even a *lot* of blood, sweat and tears and sometimes even death – gruesome death – all of which can be easily ignored for one to simply *imagine* and for which there is no cost, no risk, and no loss. The imagination is, literally, unlimited, and not encumbered by such things – which means the imagination is never completely realistic because reality imposes limits. Of course I could say "easier said than done" but let's not reduce it to a cliché – it's important to not give in to that temptation right now. So that when an engineer, trapped in his office – or perhaps more existentially – imagines someone on a rocket bound for glory – or Mars – this might be the adventure he seeks for himself, a vicarious adventure. It's easy enough for such a bored chap to think that this would be better than spending his days trapped in a cubicle, but at the same time the guy on a rocket probably doesn't feel like an adventurer, especially not after all the years of training and preparation. In fact – once things are underway – the guy on the rocket may feel quite bored to death.

Then in addition to the very major problems of BOREDOM and SECOND THOUGHTS there would exist the threat of things like stroke or heart attack. These or any other unexpected health crises could occur during the long, eight month flight or afterwards and would spell disaster and instant death to the mission. It seems there would be a baseline level of stress and anxiety that one could not alleviate, in flying in a *rocket to Mars*. In that regard alone there would be a huge advantage of an initial colony of 20 or so rather than just 3 or 4. Losing 1 guy of 4 early on would be a much bigger disaster than losing even 2 or 3 out of 20.

The bottom line is that there aren't 80,000 people who are qualified to either travel to or live on Mars, if you're considering physical, personality, and genotypic characteristics, all of which have to be considered with a degree of scrutiny and that will most assuredly not be in accordance with EEO (Equal Opportunity Employment) anti-discrimination hiring guidelines, no sir. *Genotype*, you're asking? Well, of course! These so-called astronauts will be, besides their role of astronaut, a modern version of Adam and Eve.

They will be not just the embodiment of millennia of evolution, but the apotheosis of all Humankind! Surely this should not be taken lightly! Surely we can not entrust this to the standard job application.

Given all the medical and health considerations, there needs to be a Gattaca style genetic screening of applicants – or even a subculture of eugenics. How else to minimize the threat posed by disease or inferior genotypes, merely inconvenient on Earth but that could kill our Martian future.

Fortunately, the technology exists, now, to examine an individuals complete genome, or DNA, and in only a matter of hours. So it's not too much, and is in fact completely realistic, to expect that the DNA of any and all headed to Mars will be...perfect. But what does this mean exactly, and how do you find these perfect specimens where there may be none. If we take the consideration of genotypic criteria seriously, and we have to, that narrows the field considerably, to perhaps a hundred or so finalists, worldwide. This might allow the start of a small colony – and a less absurd fantasy than 80,000. *A less absurd fantasy.*

EVEN IF WE WERE TO GO IN THE OPPOSITE DIRECTION AND THROW the doors wide open, to pre-qualify one and all, there are still perhaps only a few hundred who would be eager to go to Mars AND who satisfy some minimal set of physical, mental, and genotypic requirements. So it becomes clear that Mr. Musk's plan to send 80,000 (just in the first wave) is outrageous. There would be great limits imposed on the growth of a colony by whatever infrastructure can be built – bearing in mind we'd be starting from zero. It seems that whatever size Martian colony would consist entirely of people sent from Earth, and that rather than growing, the population would be in a constant state of decline, there being no compensation for the death rate other than a once every twenty-six month influx of new residents. It seems that a point of diminishing returns would be reached very quickly, that a colony would max out at much less than even a hundred, as a result of negative feedback and unrealized expectations (and using Antarctica and the ISS as reference points, discussed in those sections).

Despite the *profound* unlivability of Mars, there may be a certain novelty to a small colony. In one scenario – purely hypothetical, starting with a base population of 100, which is ambitious compared to the crew of 2-4 that Mars One and NASA have suggested, and a death rate of 30% between new arrivals – an influx of 100 residents every twenty-six months would provide a fixed base population of 330 within 15 years. At which point about 100 residents would die for every 100 new arrivals.

But no matter how quickly an infrastructure could be built, an influx of 80,000 people makes no sense, over any length of time. Even on Earth, developed countries can't absorb that change in population quickly without adversity.

There are so many ways to look at this. Would it be only the rich, who could pay for the ride, or the more democratic 1 in 100,000 worldwide (given the Earth's estimated 8 billion people) that Elon Musk suggests would comprise the 80,000 colonists from his announcement made a few months before the Mars One ad ran? These are two different things.

LET'S RUN THE NUMBERS AGAIN, REGARDING THE 80,000. Assuming 1 per 100 applicants qualified, that would require a base of 8 million applicants, to result in a colony of 80,000. And I would expect these 8 million, in turn, to be (at most) 1% of those targeted by recruiters (people 15 to 40 years old) – or 800 million. In other words you'd need a base population of 800 million 15 to 40 year old earthlings to end up with 80,000 trained astronauts on Mars. I'm not saying it's impossible, but building the Great Pyramid of Giza wasn't impossible either, and I'm referring to the magnitude of the recruiting effort, not that of the colony. That sounds like every 15 to 40 year old on the planet! That would be some advertising campaign.

> Our priorities are a lot different than I thought if 8 million people would consider living on Mars right now, such that that many would submit an application, even before terraforming begins, the Mars as portrayed in "Total Recall," the Arnold Schwarzenegger movie where his head practically explodes when his character is accidentally out in the open without a spacesuit – and which demonstrates how both Hollywood, and the Mars advocates, consistently downplay the unlivability of Mars.

I doubt there are 8 million people seriously interested in living on that Mars, but there may be 1 million Mars advocates (including many school kids). Now, of this 1 million, you can be sure that 99% would NOT actually consider going to Mars, but are simply intrigued by the idea of space research generally – and may even contribute millions of dollars, or start up R&D companies (that may receive federal funding

– this tends to make things worthwhile), which of course creates jobs and fuels the momentum – which then becomes an entity unto itself, around which we have media buzz and all kinds of electricity...all this, remember, while only a tiny fraction of this subculture would actually want to live on Mars – but at the same time sport the glossy decals, the slick logos, the official patches. The True Believers. If you're not among the 1 million Mars advocates, suffice to say you're a normal human being, diligently ignoring the headlines...the *electricity*. But anyway, let's say that of the 1 million Mars advocates that 1% apply for the "job," this comes out to 10,000. Then, if I stick to my first assumption, that 1 per 100 applicants qualified, i.e. were admitted to and completed training, that would be only 100 people. (And to keep the numbers simple, let's ignore the fact that the training, which would have a duration of at least a year, might possibly kill the average couch potato Mars advocate.) Total yield: one hundred people qualified to travel to Mars, emissaries of Earth, ambassadors of Humankind. Total! That's the entire colony! The entire "First Wave!"

And there would obviously be fewer applicants for each successive wave, with the numbers tapering significantly downward, since we'd be recruiting from the same base Earth population, with the first wave absorbing most of that seething, pent up "mania" that's been brewing since man first looked up. There may be no one left for the "second wave!"

But Mr. Musk's plan is to send up 80,000! How do you admit 80,000 people to go to Mars if only 100 are qualified? Well, there's one easy solution, and one not so easy. The easy solution is to ease up on the selection process – to lower the standards. That's right, I said to *lower* the standards! Maybe we don't need to worry so much about who we designate to *carry the seed of Humanity to another planet*. But hold on, like I said there is another way (although it's the not so easy way): convince Normal People – you know, the ones who've been diligently ignoring the headlines and *electricity* – that they should get on a rocket and travel to Mars even though they think the whole thing is, well, crazy.

So, Mr. Musk, good luck with your 80,000.

10

A New Vocabulary

Mars is another planet, its natural laws are completely different from those on Earth, the pictures sent back by the rovers do not convey this, and therein lies the primary illusion that Mars is like Earth.

IN SPEAKING OF MARS, WE NEED TO INVENT A NEW VOCABULARY, since even obvious reference points are, upon inspection, misleading. Mars isn't like anything on Earth, so we shouldn't use the same words to describe things on Earth to also describe things on Mars that may actually be quite different for the sake of convenience, because this results in confusion – and leads us to trick ourselves into thinking that Mars is like Earth.

For example, if we give a certain meaning to the word desert, then call something on Mars desert, it would, obviously, have the same meaning as calling something on Earth desert, but which would be a mistake if a desert on Mars is significantly different from a desert on Earth – and the two things are, in fact, significantly different, so calling Mars a desert is obviously a mistake since it leads one to think that Mars is like Earth, as if one might walk from the Sahara onto Mars without noticing a difference, whereas such a stroll would be quite, and immediately, deadly.

We generally don't live in desert environments for obvious reasons – they're hostile to human habitation – but some people have acclimated themselves to life in or near the desert so we know that deserts, as the term applies to deserts on Earth, can be, at least to some degree, "livable" – so we give this meaning to "desert." Which is why calling the surface of Mars "desert" is terribly misleading. One would easily conjure images of the Mojave or the Sahara, which are, in fact, nothing like Mars! And wouldn't that be my or anyone's natural inclination, to think that – based on pictures alone and official descriptions of Mars? Should I know that NASA is redefining "desert" without them telling me? And why would NASA redefine "desert?" That's the comparison I made early on, naturally, since the pictures of Mars and some of our deserts both have that brownish "desert" look. See how, with the help of NASA, I tricked myself into thinking Mars was like Earth! Has NASA tricked itself too? Clearly we could mitigate significant confusion by coming up with a different word entirely to describe the Martian "desert."

First, most people associate the desert with extreme heat. In fact, Mars is the opposite, and is a subfreezing cold planet – although in places it is sometimes above the freezing point for periods of the day but then returns to minus 70 F and colder every night. Some parts of the planet experience a general warming effect – with temps above freezing during the day – at certain times of the Martian year – but even here the extreme cold returns every night. Mars has been described as having, not only seasons but, in Robert Zubrin's *The Case for Mars*, "four seasons of similar relative severity to our own."

Zubrin's claim that Mars has "four seasons of similar relative severity to our own" conjures ideas of Spring, Summer, Winter, and Fall, which is unfortunate because all these words have deeply rooted connotations that, when speaking of Mars, can only mislead, and severely.

Second, on Earth, most deserts get regular rainfall (or snow), infrequent but still regular, and what infrequent rains that do fall can be quite heavy, enough to cause flooding and erosion. But there is no rain in the Martian deserts, none whatsoever. By that I mean the entire planet, since the entire planet is desert (for lack of a better word). Places that get regular rain, like deserts on Earth, should obviously not be classified in the same way as places where it *never* rains, as on Mars! To avoid this remarkable distinction, and treat the two things as if they were the same, and to promote the idea that the two are the same – such that it misleads one to think that a place that might be called a true desert because it gets absolutely no rain *is habitable* – is propaganda. And by rain I mean water rain, not the other "rain" that scientists sometimes use for the sake of expediency to describe other things falling from the sky when talking about other planets, like carbon dioxide or ammonia – a distinction that needs to be emphasized and writ large if one is seriously considering *living on Mars*.

You can't be on Mars (which is all desert, or "desert-like") without quickly dying. You would asphyxiate, then your body would be freeze-dried. The outrageous fact is that pictures sent back by the rovers do not convey this reality – pictures sometimes DO lie, and this is a large part of the Mars advocate propaganda. There's a huge perception issue with these pictures, many depicting what appears to be an adorable Arizona sunset but where the temperature is actually about minus 70 degrees F, that NASA seems to downright enjoy foisting upon us, and in this they are downright irresponsible, along with the science writers.

Mars is another planet, its natural laws are completely different from those on Earth, the pictures sent back by the rovers do not convey this, and therein lies the primary illusion that Mars is like Earth, at least from the standpoint of living on it. An illusion that has taken hold of many scientists, science teachers, and students alike and quite set their minds afire and the same for opportunists who put ideas of "space tourism" into their heads. It would make as much sense to build a stairway to heaven.

There do, of course, based on photographs, appear to be visible similarities between the Martian surface and some of our deserts, but there's a lot more to a desert than meets the eye – than can be conveyed in a humble photograph – and this is certainly true of the Martian "desert." As with oceans and lakes, you couldn't tell from photographs alone that oceans are saltwater, while lakes, usually, are freshwater – there's a big difference. Paint makers have many names for their different shades of white, but we (the English speaking world) don't have different names for different types of desert. Only if you're buying white paint do the creative and fanciful names of the particular shades of white matter to you, likewise most of us don't live in or near a desert so we just don't bother to distinguish between – to name – the various types of desert. But there's enough difference between the so-called desert of Mars and Earth's deserts for it to matter – especially if you plan on moving there. Because the experience of living on the "desert" of Mars would be completely different from any of those on Earth, and not just in terms of temperature and precipitation, but also of air and air pressure, which I describe in detail elsewhere.

The Martian "desert" isn't a desert in the same way that a hurricane isn't a tornado, which in turn isn't a "dust devil" (a type of tornado that occurs on Mars), a word thoughtfully coined by NASA obviously to distinguish it from hurricanes or tornadoes, since they're (allegedly) not as dangerous as the other two and NASA wants to make that clear, to downplay the risk denoted by hurricane and tornado and that might otherwise hinder continued research and cause them to lose their jobs.

SELF-SUSTAINING

Let's look at this idea that a colony on Mars might be "self-sustaining." When a Mars advocate says that a Martian colony can be self-sustaining, this is another example of how the English language itself falls short when discussing Mars. Because surely he doesn't mean self-sustaining in the familiar sense, such that something doesn't require a lot of maintenance, or doesn't require one to check on something with diligent frequency, or that something would exist through some *autonomous and invincible* force of nature like a tree growing in a forest. What a Mars advocate means by self-sustaining is more like someone in a coma is self-sustaining if they're lying in an ICU unit with 24-hour care, grotesquely wired and intubated – which would, more or less, be the state of things on Mars – but which connotes quite the opposite of "self-sustaining!" In other words, a colony on Mars, in the *ordinary* sense of things, would be "unsustainable." So you see how the Mars advocate turns upside-down ordinary ways of thinking – which is another form of propaganda. More doublespeak. Also "in the ordinary sense of things," one might think my comparison to a coma patient is sarcastic – but it isn't! It might be if this discussion was confined to the "ordinary sense of things" but when speaking of Mars – and comparing it to Earth – we are not confined to the ordinary sense of things – let that be perfectly clear. You would therefore be mistaken to think that my comparison to a coma patient is meant in anything but the utmost sincerity.

If a colony on Mars were to be truly self-sustaining, it would have no ties to Earth whatsoever, and no one on Earth would need to be aware of it, much less guide, assist, or direct its development, or in any way control it. Its existence would be *autonomous and invincible* – like humans on Earth. But the prospect of this, in the near or long term, is laughable. Martians certainly wouldn't be autonomous if there were any dependency on trade or delivery of goods. In the same way that people in some third world countries are dying of starvation – right now! – but we are oblivious to the fact, we would be no more aware of people on Mars dying of starvation – or eating each other. That's "self-sustaining!" You would no more care about Mars and Martians than the price of tea in China – on that day, a Martian colony would be "self-sustaining." Even NASA would say "what's Mars?" If there are creatures living on Jupiter right now...that's self-sustaining!

There would never be a large colony (80,000 for example) on Mars but this will *not* be due to a lack of sustainability. How simpleminded are the Mars advocates? – this is why we call them "space cadets" and I don't mean in the mildly derogatory sense.

> **Whether or not sustainability is possible isn't even the issue – because of all the intermediate steps between zero and sustainable that are countless, unaccounted, and unaccountable – and where the goal of sustainability could only be a dream.**

The reason we're here on Earth isn't because of anything being sustainable. The reason I exist here and now can't be reduced to some childish notion that I'm sustainable. This is more of that engineering lunkhead thinking that's confused for intelligence. Do you want to live in Antarctica? Do 80,000 people live in Antarctica? Do any scientists live in Antarctica as permanent residents? No No and No. Because life in Antarctica is not sustainable. How then do you expect where things are MUCH WORSE to be *sustainable*? – even if getting there was EASY?

MARS NOT FOR ADVENTURERS

People enjoy being outdoors, especially people who are adventur-ous! Adventure can be *equated* to the outdoors, but it seems this could never be the case on Mars, where everyone would be trapped for life in an airless habitat, on a planet where the only winds that blow are not cruel, but deadly – what adventure seeker wouldn't want that? So the Mars advocates need to entirely rethink this "Mars is for adventurers" idea, and quickly.

11

Wikipedia

〜〜〜

"Hospitable" suggests a very specific refinement of living conditions, with very particular connotations, such as grandma's house on Thanksgiving day, providing conditions that afford not just comfort, but the epitome of pure, sensuous, carefree comfort – which obviously could not possibly describe Mars on its best day.

WIKIPEDIA IS A VAST, INTERNET-BASED ENCYCLOPEDIA, AND presents things in a scholarly, matter of fact way, with a generally high degree of consistency. There are versions for 286 or so languages and is easy to use, is widely used, and so is a great way to disseminate information. Something important to keep in mind when using Wikipedia is that the articles and definitions can be edited by anyone – including those who wish to promote an ideology – or agenda – even though this is generally contrary to the terms of use or sense of objectivity and fairness that "more or less" direct its editorial policy. Many have become wary of Wikipedia for this reason – nonetheless, Wikipedia is hugely popular and highly trusted – which makes it both a powerful and convenient propaganda tool of the Mars advocates.

There is, of course, an entry for "Colonization of Mars," but which seems to ignore Wikipedia's editorial standards and "neutral point of view," in particular a statement that describes Mars as being "hospitable," which occurs (or did at one time) at the beginning of the article, where it is more likely to be read. This alone turns the entire article into a propaganda piece – in which case "Colonization of Mars" shouldn't exist in Wikipedia at all. In case you didn't know, Mars is not warm and inviting! – in either the relative or absolute sense – as if accommodations could be set up right now on Mars that are equivalent to a bed and breakfast in Santa Barbara. Even for the most devoted Mars advocate, to claim that Mars is "hospitable" isn't just ridiculous, but insane. It's an example of *doublespeak*, which means to call something its opposite, and commonly used by propagandists.

It is propaganda in the fullest, Orwellian sense. On the ridiculous pyramid, claiming that Mars is "hospitable" is the apex – whether in Wikipedia or anywhere else.

Whereas it is simply misguided for the disinterested individual to think that Mars is hospitable, whose exposure may be limited to an accidental glance at a NASA photo – such as might occur while watching the evening news – it is extremely *delusional* for a student of the subject, or specialist, one who has the confidence to edit a scientific Wikipedia article, to insist that it is so. Hospitable means warm and inviting – something Mars clearly is not based on even a casual review of the basic facts.

It would be difficult to defend the author or authors of the Wikipedia article "Colonization of Mars" by saying they probably intended to say "inhospitable," because that simply means not warm and inviting, which would completely miss the point, because based on all evidence, Mars is threatening, hostile, and deadly – and this does not equate simply to "inhospitable." On the other hand, did the writer of this part of the Wikipedia article mean the more general "habitable?" Even that would be an enormous exaggeration for a place that is, again, and based on the known facts, threatening, hostile, and deadly – and considering that even Antarctica is usually considered uninhabitable, and a well known reference point of the Mars advocates. These two words are quite different – habitable and hospitable – a place could easily be considered habitable without being at the same time hospitable. Antarctica could be considered habitable (although most references to it do not agree even with this) while at the same time it is obviously not hospitable. San

Diego is hospitable, Siberia is not, although they are both habitable. I think you get the idea. Again, these are completely different words, with completely different denotations and connotations.

> **This means that whoever wrote this part of the Wikipedia page for "Colonization of Mars" was either a victim of the Mars advocate propaganda, or one of its sources. If there's a mass delusion regarding Mars as being "hospitable" (and therefore also habitable), given the high exposure and respectability of Wikipedia, it's no wonder!**

But let's take it a step further: hospitable to me, and probably also to you, suggests a very specific refinement of living conditions, with very particular connotations, such as grandma's house on Thanksgiving day, or a bed and breakfast in Santa Barbara, providing conditions that afford not just comfort, but an extreme degree – the highest degree conceivable – the epitome! – of pure, sensuous, carefree comfort – which obviously could not possibly describe Mars on its best day. In fact, in that light, one is hard pressed to describe even Earth as being hospitable! But Mars?! In any case, let's agree that one must avoid leading others to think, in some scientific sense, that Mars is "warm and inviting," especially if it could lead anyone to actually move to Mars! – or to think they could.

I edited the Wikipedia article, to remove the word "hospitable." A few weeks later, hospitable was back in the article. Do you see the problem? Given the precision of the Wikipedia article generally (aside from its accuracy), this can hardly be considered an oversight. Wikipedia policies and guidelines might even consider this to be an act of *vandalism*, since it seems to be the "deliberate addition of plausible but false information" to an article – although in this case it seems to go way beyond that. I will say it again:

> **It is delusional – in the true, psychiatric sense – for anyone, scientist or non-scientist, to insist, as some editors of the "Colonization of Mars" article have done, and which is evidence that this delusion exists among at least some Mars advocates – that Mars is "hospitable."**

Yet this is a recurring theme of the Mars advocates. Some Mars advocates are probably full blown psychotics – which is not to suggest that any of the more entrepreneurial Mars advocates are psychotic – or that they're not. This should not be taken lightly – many people classified as mentally ill have just this sort of "belief system."

But for the sake of humanity, the exploration of space needs to be guided by hard-edged sanity, not delusional hopes and dreams – or psychos.

More generally, the problem with the "Colonization of Mars," as presented in Wikipedia, is that it treats the whole thing as if Mars is already colonized – which it obviously isn't. Anyone – school students or old-timers alike – who was learning about the planets for the first time could easily be convinced otherwise, that, for example, hydroponics was already being used to feed people living on Mars, as depicted by the artistic renderings appearing throughout the Wikipedia page for "Colonization of Mars." Which is of course completely misleading. And science-fiction. Wikipedia was never intended to be a repository of every single known thing, idea, or theory of fact or fiction. To suggest, using a widely published source such as Wikipedia, that man is already on Mars, or will inevitably colonize Mars, isn't just misleading, it's propaganda. Wikipedia, despite its wide ranging scope, isn't the encyclopedia of science-fiction and fantasy, nor a venue for prognosticators or fortune tellers. Encyclopedias, including Wikipedia and other respected sources, inform as to the way things are and known to be, and not through the dreams of science-fiction junkies.

Which is not to say that there should be no mention whatsoever in Wikipedia for the "Colonization of Mars." But with something realistic and free of promotion or fantastical embellishment, as per the editorial policies defined by Wikipedia, such as "The colonization of Mars refers to the theoretical and science-fiction inspired idea of man living on Mars." That would acknowledge the idea of colonizing Mars, without creating the distinct impression that man already lives on Mars, which is propaganda. Considering that man has never been on the planet Mars, it goes without saying that there are no colonies on Mars, and furthermore it's not unreasonable to speculate that Mars may never be colonized – or even visited by man – such a thing is certainly *not* inevitable and for legitimate technological – and non-technological – reasons. In fact, such a thing – man travelling to and colonizing Mars – is not possible in this day and age – which means impossible. That said, it doesn't make sense that the entry for "Colonization of Mars" would be so massive and full of detail, which belies the reality. Or that the illustration depicting hydroponic farming should be made *more realistic*, as one Wikipedia editor had suggested. Statements of scientific fact on the "Colonization of Mars" Wikipedia page might be perfectly suited to the "Mars" Wikipedia page, which would at least make it easier to research Mars as a planet – about which there is much to be said – without being distracted, confused, or lead astray by the bizarre fantasies being promoted by the Mars advocates.

One would at least expect the "Colonization of Mars" page to describe, if not human colonies that are on Mars (because there are none) then characteristics of Mars that could be expected to lend themselves to human colonies...but there are none – because the facts suggest that Mars is, in every sense of the word, unearthly – and to a degree that is shocking – so what can one do but speculate about the prospects of a human colony, and the measures man would need to take to protect himself from Mars – and how crazy the whole thing is? But speculation is not allowed on Wikipedia. Speculation is treated as vandalism by Wikipedia – even when there is nothing to do but speculate. So the existence of "Colonization of Mars" on Wikipedia is a paradox. Which begs the question, why is there a "Colonization of Mars" article at all on Wikipedia? Propaganda, what else?

Protect himself from Mars – which is an understatement of biblical proportions. To be sure, humans going to Mars is like angels coming to Earth with their wings ripped off. But of course you wouldn't find *that* in Wikipedia. Unless someone at NASA said it but the pyramids will crumble first.

THE WIKIPEDIA EDITS

2/15/13 – I replaced the existing introductory paragraph for the Wikipedia page "Colonization of Mars" with the following:

> The "colonization of Mars" refers to the theoretical and science-fiction inspired idea of man living on Mars.

This contains no promotional embellishments or description of Mars as being "hospitable." There was, and remains, much more than that on the web page, with elaborate details provided by the Wikipedia community – the top portion of each Wikipedia page is for the most basic of introductions, especially when the entries are lengthy, to appeal to a broad audience. Since Mars isn't colonized, anything more elaborate than the above might give the opposite impression. And since it seemed that the "Colonization of Mars" Wikipedia page was already being used as a form of propaganda, I knew I could only go so far in my attempt to keep things rational, that if I said something as blunt as "Mars is dead and deadly" the Mars advocates would swoop in – but it didn't take much time for them to swoop in anyway and restore their propaganda.

3/29/13 – A few weeks later, someone edited this [Cardiomals], deleting my portion, but expanded upon it, and added about three sentences, but without significantly changing the meaning. I only discovered this months later while investigating the matter for this book. I'm not a frequent editor of Wikipedia and don't return to see if someone has edited my contributions, since anyone can edit Wikipedia and no one is entitled to have the last word.

4/1/13 – Within two days of the edit made on 3/29/13, someone else [Againme] added a short paragraph, to the same section of the Wikipedia article, at the top of the page, including this sentence:

> It is the focus of serious study because surface conditions, such as the availability of frozen ground water, make it the most hospitable planet in the solar system other than Earth.

Are you in shock? Hospitable! You wouldn't say "make it the most San Diego-ish planet in the solar system other than Earth," would you? Someone wants to make sure that the world thinks Mars is "hospitable!" But of course it was April first, which is April Fools Day! What perfect timing!

Use of the word "hospitable" in any description of Mars is misleading at best. Substitution of the portion of the Wikipedia page that I added, which did not include the word "hospitable," by another with material that contained the word "hospitable" demonstrates how aggressive the Mars advocates are, how unreasonable, how irrational they are, and is evidence that they employ propaganda, and effectively. The question is, why are they doing this?

Shortly after this I was blocked from making further edits to Wikipedia – due of course to the Mars advocate sentinels – standing guard at the gates of perhaps the world's biggest propaganda machine and taking no prisoners.[1]

---

1   It may be important to note that the Wikipedia editor named above, "Againme," is not me, in case you are led to think otherwise.

12

The Case for Mars?

Colonizing Mars is not intended to be a stunt, like the Moon walk, or the rhetoric of entranced idealists (overeducated space cadets) – although that's what it is – but the glorious next step in man's evolution! – if one is to take the Mars advocates seriously.

THE THESIS OF THE MARS ADVOCATE IDEOLOGY IS THAT MAN should – must! – colonize Mars. Within the main body of this ideology – as represented in books and other media – there exist many fallacies. Not all of them are obvious. They range from simple to complicated, from contemptible to laughable. Some of the ideas may seem credible (especially to those employed in relevant industries or otherwise on the bandwagon), but all are fallacies nonetheless, and all deserve to be fully scrutinized – and ridiculed, considering the fact that humanity itself, and the future of Earth, may be at stake. A book by Robert Zubrin, *The Case for Mars*, contains many of these fallacies – in fact may be considered a guidebook to Martian fallacies. It may in fact represent the epitome of the Mars advocate mentality – in terms of the potential damage it could do to man's future.

Clearly *The Case For Mars*, which promotes a human colony on Mars, might never have been published were it not for the authors status as an ex-NASA engineer and other professional associations, such as with renown science-fiction writer Arthur Clark, author of *2001: A Space Odyssey*, who Zubrin convinced to write the forward to his book *The*

Case For Mars – although it seems the endorsement of a science-fiction writer would, by itself, undermine whatever validity Zubrin's ideas may have.

It would make more sense if I wrote Zubrin's book and he wrote mine, since any expert on the subject (Zubrin) would be *forced* to conclude that man can't live on Mars, other than as a ghastly, weird experiment.

> **Man can't live on Mars, other than as a ghastly, weird experiment, like pulling the wings off a fly while proclaiming "look, it's still alive!" in the moments before it dies.**

Zubrin's *The Case for Mars* may indeed be nothing but a compilation of stray notes and idle thoughts from his days at NASA and driven by the publish or perish mentality and enabled by a publishing industry more and more hell-bent on scandal and junk. It is a masterpiece of propaganda, which by itself is undeniably fascinating. So let's examine the fallacies, break them down and lay out the parts so everything is plain to see.

First, Zubrin never fully explains the moral imperative behind colonizing Mars that he and others claim to exist – whereas the imperative to be on Earth is both obvious and imminent. Not a single word in his book makes the prospect of going to Mars seem in any way attractive – or even doable. Zubrin doesn't explain why we NEED to get to Mars. His book may serve as a technical reference – or as background for a sci-fi novel – but his science and crackpot theories are more disturbing than convincing. He doesn't convincingly explain the ethical or practical imperative in moving to Mars, or his sense of urgency.

If we had to colonize Mars, *The Case for Mars* could very well provide the blueprint, but in the same way that plotting the most perfect course for climbing Everest doesn't create a reason to climb Everest, a blueprint to colonize Mars doesn't create a reason to go to Mars. Taking it a step further, if I had a map to Hell, I'd have the good sense to DE-STROY IT! Go to Hell, go to Mars – there's no difference!

In his book, Zubrin states "There are real and vital reasons why we should venture to Mars. It is the key to unlocking the secret of life in the universe...reaffirm the nature of our society as a nation of pioneers...a spacefaring species with no limits to its resources or aspirations..."

This is the language of the propagandist, and a deranged premise upon which to base the enslavement of future generations of human beings on a desert planet with no oxygen or food sources. The idea of enslaving future generations of humans on some desert planet is absolutely outrageous, as if those born on Mars wouldn't have free will and come to Earth as soon as they could, which obviously has more potential than Mars but which the misguided Mars advocates – especially current and ex-NASA space junkies like Zubrin – take for granted.

ONE OF ZUBRIN'S MORE OUTRAGEOUS CLAIMS IS THAT THERE WOULD be trade between Earth and Mars. This idea isn't just ridiculous, it's superfluous. If there is an undeniable need to either explore or colonize Mars, and that it is already habitable, as the Mars advocates insist, then issues concerning returning to and trading with Earth evaporate. One need not imagine rockets lifting off from Mars and returning to Earth, and certainly a colony on Mars should not be delayed by the lack of ability to do this. Why would people be returning to Earth? That would be like souls in heaven...returning to Earth! – which is, of course, absurd! The idea of people traveling "back and forth" is completely superfluous, in fact a two-way highway between Earth and Mars seems to flatly contradict the notion of a self-sustaining colony. The same is true for any kind of trade or commerce between the two planets – this simply wouldn't need to be realized to justify a colony on Mars. So why is NASA and the Mars advocates hung up on this, the current inability to fly from Mars back to Earth?

For example, Zubrin has suggested that *deuterium* be shipped here from Mars. The ridiculousness of this idea speaks for itself – there might not be a better example of the sociopathic engineer idiot savantist mentality.

That Zubrin would choose deuterium as the premiere example of this outrageous conception of interplanetary commerce is just another example of his multifaceted madness and by itself practically undermines his entire thesis.

It's just more crazy talk and completely outside the realm of intelligence, with absolutely no scientific reasoning and absolutely cost prohibitive, to say the least. Deuterium, also known as heavy hydrogen and which comprises heavy water, is an isotope of hydrogen that is found in regular water, from which it is easily (more easily than *shipping from Mars*) condensed through electrolysis and from which we are able to obtain large quantities; more than one atom per ten thousand hydrogen atoms has a deuterium nucleus. [http://www.sciencedaily.com/ releases/2009/05/090511181356.htm]. It is used for nuclear and medical research and exists in such abundance that we will never run out of it so there would be absolutely no reason to ship deuterium from Mars even if this could be done *for free*. That this would be presented as a "strong argument" by such a highly educated and prominent Mars advocate demonstrates the profound weakness of their general thesis, so as you can see there is simply NO END to the propaganda the Mars fanatics are pitching, like snake oil.

Mining gold or other rare-earth elements on Mars to bring back to Earth would be cost prohibitive and not worthwhile in any other sense, in light of the success of our species (which is, of course, what we're talking about) – we've gotten this far without trade with Mars, so why now?

That's the great thing about Earth, it lacks for nothing – everything we need exists in extravagant excess. Mindblowing, self-sustaining excess. And fortunately, man has the intelligence to use her resources wisely. Right?

There may likely never be rockets launching from Mars to Earth, but you can be assured they would not be carrying deuterium (and every tank of deuterium means one less person desperate to get from Mars back to Earth).

Perhaps Zubrin has built a fountain of youth that requires deuterium from Mars!

It's almost as if Zubrin has written a script for a movie, where special effects will make anything seem possible. Because in real life, the shipping of deuterium from Mars to Earth, and most of the other ideas he's suggested, would and could never occur – and for entirely practical reasons, not mental hangups as Zubrin repeatedly but ingenuously claims as the reason we're not *already* on Mars. *The unfeasibility of any kind of shipping from Mars to Earth can not be overstated.*

> **Zubrin states** "the spaceship crew faces risks, but it's not as if there are enemy armies trying to kill them."

More strange brainwashing logic. In fact, the vacuum of interplanetary space is far more hostile and unforgiving than any army known by man.

> **Zubrin states** that the spaceship crew has the "knowledge that as soon as they get back to Earth, their fortunes are made, they will be golden people, celebrated as heroes."

This isn't just ridiculous – Zubrin punches himself in his own Mars Advocate face. As I say elsewhere, they can celebrate themselves as they throw back calcium supplements. It doesn't make sense to think that ones satisfaction in going to Mars would be in the least way dependent on "getting back to Earth." That would violate the premise of going to Mars in the first place – to create a "self-sustaining" colony – and eventually an Earth substitute. Why would anyone leave a self-sustaining colony – which seems to contradict the idea of self-sustaining? Physiological and biological changes to individuals that would either occur naturally or be devised as a form of adaptation would make returning to Earth literally impossible anyway. Colonizing Mars is not intended to be a stunt, like the Moon walk, or the rhetoric of entranced idealists (overeducated space cadets) – although that's what it is – but the glorious next step in man's evolution! – if one is to take the Mars advocates seriously. In which case every single departure of a Martian back to Earth would be an act of *defeat* – and not victory. All this means that any "reward" would be gained from simply...being on Mars! – and free of the "adversities" and apocalyptic visions tormenting Earthlings. Indeed. If people go to Mars with the hopes or anticipation of returning to Earth it would be in total violation of the basic philosophy of colonizing

Mars, and subverts the whole thing to a childish game. Going to Mars may involve a commitment that even the Mars advocates do not yet fully appreciate!

In point of fact, no astronauts made a fortune after returning to Earth. Neil Armstrong taught at a college in Cincinnati, Ohio before retiring, and absolutely shunned celebrity. Getting rich has never been an incentive for astronauts. Most start in the military, which is decidedly not a route to riches. The trip to Mars would be one way, according to almost all descriptions, and not consistent with "fortunes on Earth," which is hardly the incentive for exploration anyway. A few rocket makers might get rich, and whatever associated industries, but not the people going to Mars. So there is NO CASE to be made for the "fortune seeking" Martian wannabe whatsoever, it's just another of Zubrin's many weird ideas and mixed-up rhetoric. It's almost as if Zubrin came up with a million and one reasons to go to Mars – then picked from the bottom of the list. It might have been Zubrin's obsession with colonizing Mars quickly that spurred his departure from NASA...well, of course that was it! – and he would never get rich confined to his NASA cubicle! – that's for sure. Whereas NASA's objectives might be more conservative, to send up unmanned probes, and maybe later a few guys to scout around and then return – like with the Moon.

> **Zubrin**: Regarding the concern that we won't have enough to do on Mars, the first settlers will be "heavily occupied conducting broad-ranging field exploration."

To Zubrin, if not all the Mars advocates, it's a foregone conclusion that people exist for no reason other than to WORK; that people are nothing but DRONES serving no purpose other than LABOR. Any similarities of my grammar to radio host Alex Jones' "prison planet" rhetoric are completely unintended – though entirely appropriate. But there's an even more fundamental problem with Zubrin's expectations, especially regarding the "first settlers." There are simply too many survivability issues to think that the first people on Mars would be leisurely going about "exploring," as if Mars was nothing but a movie set, with full catering services. And what would they be investigating – whether or not the soil can support plant life? – and similar basic questions? We already have the answers to such basic questions! It would seem that the first "settlers" would arrive at a *later* stage, only after several decades – centuries! – of so-called terraforming. Terraforming would need to precede "settling," and the rovers and satellites are already doing broad-range field exploration. All of which could be preceded by a few men going up to scout around – and to truly appreciate – finally! – the grim reality that NASA and the Mars advocates have been denying all along – as they brainwash themselves and the public. But "settling" Mars? It's a stretch to say that Antarctica is settled, limited as it is to a dozen or so "base camps." With this in mind, it seems only too reasonable to predict that Mars could never be "settled." In which case it is simply incongruous for Zubrin to say that the first settlers will be "heavily occupied conducting broad-ranging field exploration."

Not to be completely sidetracked, but even if Mars was an exact duplicate of Earth, with clones of all the cities and in sync with Earth, it would still be enormously expensive to get there, the trip so boring as to be intolerable and taking eight months out of your life and, as always, extremely dangerous, with the odds of surviving the trip about 1 in 10 or worse – which is pretty bad. These would all be remarkable disincentives, even for a Martian Earth clone. There's no telling just how much *better* an Earth clone would need to be to justify such a move *even if it's the next planet over* – or how bad off you must be to think such a trip would be worthwhile. It just can't be overstated how stupid and ridiculous it is to think Mars is livable or that anyone should go there.

> **Zubrin** proposes that America bear the entire cost. "... it's a sum America can easily afford" and "exploring Mars requires no miraculous new technologies."

These are outrageous lies and more propaganda. Just getting to Mars would require several miracles, including the technology to land a heavy, cargo carrying spaceship gently – which is currently impossible.

Zubrin: Mars has "four seasons of similar relative severity to our own."

This is one of Zubrin's excellent brainwashing tricks. Mars experiences what some describe as summer and winter but so does Antarctica! On Mars these cycles do not lend themselves to human habitation. The four seasons on Earth correspond to life cycles – the growing season – whereas *none* of the "seasons" on Mars would lend themself to a growing season, a concept that may be meaningless on Mars due to the subfreezing temps that occur on a *daily basis* year-round planet-wide and which would probably continue even after centuries of terraforming. It takes Mars 687 days to orbit the sun. By analogy, this might suggest a growing season that is twice as long as the one on Earth but that occurs only once every two Earth years – *but this is really not analogous to anything on Earth because on Mars it's freezing every day!* Any agriculture on Mars could obviously not rely on such "seasonal variations" if a steady food supply is the objective – which of course it would be – and which by itself may render the idea of "four seasons" quaint and meaningless.

In fact hourly fluctuations in temperature are so extreme on Mars that the time of day would be more relevant to human activity and productivity than the time of year or season.

Zubrin: "...there is every reason to believe that geothermal heat sources could be maintaining hot liquid reservoirs beneath the Martian surface today."

There is no data to support this wishful thinking. Even if Mars was hollow and filled with water, tapping into this would expose it to a very rapid evaporation. It would be completely gone in a few hundred years even with elaborate water conservation scenarios that included airtight containment facilities. Water on Mars would be more precious than oil on Earth, you'd have "water paranoia" and vicious water wars, severely limiting the human population and any chances of a thriving human civilization on Mars.

> **Zubrin:** "...atmosphere thick enough to shield its surface against solar flares..."

This is absolutely false and more propaganda. The Martian atmosphere is known to be less than 1% that of Earth and does NOT provide a good barrier to UV radiation, much less solar flares. The radiation that a person would be exposed to at the Martian surface is about 20 mrem/day, with daily spikes of 1,000 mrem, compared to about 0.8 mrem/day on Earth. You move to Mars, you're getting cancer if you don't die of radiation poisoning first – unless you're inside or underground all day.

NATIONALISTIC PRIDE

> **Zubrin** claims a mission to Mars would boost nationalistic pride.

The problem in arguing with this is, the minute you attach patriotism to something, it seems *unpatriotic* to argue the point, and which makes it an effective tool for the Mars advocate propagandists.

But of course any mission to Mars has nothing to do with nationalistic pride or patriotism – certainly not when other countries besides ours are involved. How the space program and nationalistic pride became attached to each other I'll never know.

> **Would full employment boost nationalistic pride? Free health care?** Everyone would benefit from full employment or free health care, and not in some vague, symbolic way, while only a few directly benefit from the space program, which is extremely elitist.

There are many things that might be done here on Earth that would be practical and do more than just "promote nationalistic pride," including "terraforming" the western desert states so they're more habitable and better suited to agriculture, and likewise the other deserts on Earth. There are many areas of our planet not used – terraform it! Control weather patterns to reduce storm damage! Make foods healthier to wipe out obesity! These three things, just off the top of my head, would benefit *millions* directly, unlike the jobs program at NASA which only benefits a few thousand engineers and technicians and from which there may be some *derivative* technologies but not necessarily. We need to recruit kids to study biochemistry and cell biology to discover wonderful new cures for cancer and other diseases, and longevity and anti-aging, just think of the strides that are *inevitable* (if you don't mind my indulging in a little bit of Zubrin's own rhetoric) – the technology exists to do these things NOW!

INSTEAD OF MARS MADNESS, LET'S CREATE EARTH MADNESS, A master plan to recreate Earth as a Garden of Eden and which will bring on an epoch of "rejuvenation" (to borrow again from the Zubrin lexicon) – or at least spark an interest in our young people to take up careers in science! For anyone who's got the terraforming bug, terraform Earth![1] Just give everyone at the NASA jobs program a pay raise – that should bolster the nationalistic pride Zubrin has in mind – and which has an immediate payoff!

Zubrin declares that the problems (what anyone else would call "deal breakers") concerning radiation, gravity, human factors, dust storms and back contamination are merely "monsters of the mind," like dragons and Cyclops. These and similar sentiments clearly indicate that Zubrin is off his rocker.

> **Zubrin:** "Mars' polar axis is tilted, which causes the southern hemisphere to have more extreme seasonal temperature variations."

Yes, but Mars' eccentric orbit further compounds whatever seasonal temperature variations that occur. And Zubrin seems unaware that computer simulations indicate that Jupiter's gravity has repeatedly altered the polar axis of Mars, which could adversely affect the long term prospects of Earthlings on Mars. In Zubrin's defense, this fact may have been discovered since Zubrin's 2011 updated republication of *The Case*

---

1   Terraforming is a provocative "high concept" based in science-fiction but is unlikely to ever be implemented as imagined and whose goal is to convert Mars into something more Earth-like.

for Mars (and upon which my analysis is based).

> **Zubrin**, regarding the dust storms that are known to occur
> on Mars that can cover the entire planet: "To an observer on
> the surface the surrounding area is not fogged out."

How would Zubrin know this? Zubrin completely dismisses these
massive dust storms as being nothing – even though he, or anyone, has
never experienced one. One can't know from miles high in space (the
height of an orbiting satellite) what the ground visibility is on Earth
to someone on the ground when it's snowing, and likewise the dust
storms on Mars. But being "fogged out" may be nothing compared to
the dust that would accumulate on everything, and its potentially toxic
constituents. Even if the dust storms aren't deadly, the extent to which
they might disrupt business as usual to a future colony is completely
unknown, can't be predicted, even from storm to storm, and would be a
huge wild card.

Back contamination

> **Zubrin**: "The Earth is already, now, exposed to Mars'
> surface organisms...as meteorites (SNC meteorites)...Earth
> is torpedo alley regarding rocks blasted from Mars."

The constituents of meteorites wouldn't readily get into Earth's
biosphere, so this form of contamination clearly doesn't pose a threat
to humanity. But this doesn't mean that these same constituents, delivered in some way other than as a meteor travelling through space for
perhaps millions of years, couldn't pose a threat, such as on the surface
of a spacesuit or via a human vector.

> **Zubrin**: "Martian microorganisms can't compete with
> Earth's, and vice versa."

Human germs would enter the Martian ecosystem and interact with
it in unpredictable ways. Such microbes could get back into a human

and do unpredictable, nasty things.

Zubrin states "No one will ever die of a Martian disease."

This prediction is not reasonable, and more brainwashing. But certainly no one will ever die of a Martian disease *if no one goes to Mars.*

Zubrin states rather explicitly that the microbiology of Mars is mutually exclusive of the microbiology on Earth. But in the same paragraph – the same paragraph – he then states that "Martian life may lend itself to cures of Earth's diseases." Both of these statements – in the same paragraph! – obviously can't be true. His illogic is dazzling. This demonstrates the arrogance of Zubrin (and many engineers): they think they are above even the rules of simple logic and reason, with their ridiculous claims and predictions. But more importantly, this also demonstrates the problem with rocket scientists in particular thinking they should have a say in whether or not man should colonize Mars.

INCENTIVES TO GO

Zubrin compares the hardships that would be experienced by the first settlers to those experienced by soldiers in World War II, which is just one of a multitude of his apples to oranges comparisons. It seems Zubrin forsakes the rigors of logic and rhetorical analysis for convenience over and over again. Which of course weakens his overall thesis.

First, WWII had a payoff (the end of Nazism), and the payoff was imminent, the ethos of the soldier was not just accepted but admired, and the payoff worthwhile: the freedom of millions, the risks necessary and easy to understand. None of this is true in settling Mars, and has no analogy to WWII. Also, a war machine is decidedly more manageable and more reliable than the machinery of space travel - guys can get their hands on guns and there's an immediate payoff to killing the enemy. You could even say that war is an aspect of human nature, and familiar. There has always been a war going on somewhere on Earth. Heck, people like wars. On the other hand, Mars is strange and deadly. There's no obvious need or immediate payoff in going to Mars.

Zubrin describes the hardships man has had to endure through the ages, that therefore, by comparison, flying to and living on Mars should be easy - the kind of logic one might use to brainwash someone, as if shame or guilt were an appropriate motivator for something like this; as if progress is a hard road and no one is excused. It does actually inspire me a little – but not to go to Mars.

A lot of ideas proposed by Zubrin include a component, reading between the lines, known as "crossing your fingers." He makes critical assumptions and glosses over big issues. For example, the building of brick structures on Mars. It seems the entire future of life on Mars would be characterized by brick structures set into the Martian landscape. But it seems an existence confined to this would be demoralizing and not adventurous.

Zubrin insists Mars could support plastics, ceramics, and glass industries. The need for abundant amounts of water for all these suggests otherwise. Without convenient access to an abundance of water, life on Mars would be an ongoing drag, and Zubrin proposes no way around this, where one has to extract water either from the atmosphere or by heating dredged up ground material sufficient just for a Mars "outpost." Very discouraging. I can hardly imagine a thriving Martian society.

Zubrin explains how protein could be obtained from mushrooms and tilapia fish. Fruit trees would provide, besides fruit, wood, which could in turn be used for furniture or rendered into plastics.

These things suggest a more sophisticated existence than I would have imagined, and raging optimism on the part of Zubrin – but not an existence I would want, under any circumstances, for any amount of money, now or ever.

There is a particular paragraph, on page 253 of Zubrin's *The Case for Mars*, that is noteworthy, because it contains several ideas juxtaposed quite conveniently for me to make a devastating point. Here is the paragraph:

"This forced pragmatism will give Mars an enormous advantage in competing with the less stressed and therefore
more tradition bound society remaining behind on Earth.
A frontier society based on technological excellence and
pragmatism...will perforce be a hotbed of invention, and
these inventions will not only serve the needs of Mars
but of the terrestrial population as well [meaning Earth].
Therefore, they will bring income to Mars (via terrestrial
licensing) while at the same time they disrupt the labor
rich terrestrial society's inherent tendency toward stag-
nation. This process of rejuvenation will ultimately be the
greatest benefit that the colonization of Mars will offer
Earth. And it will be those terrestrial societies who have
the closest social, cultural, linguistic, and economic links
with the Martians who will benefit the most."

Even as fiction this would be difficult to stomach. But this is not fiction, it is meant seriously – in fact, the above excerpt is brainwashing at
its finest. It's obviously very high-minded, but let's pick it apart. First,
"A frontier society based on technological excellence" is incongruous.
Frontier societies (the classic example being America's "wild west") – as
would certainly be the case for people on Mars, even after the initial
stages – are based on brute force survival, which is not exactly an environment of "technological excellence." There is absolutely no historical
precedent where a "frontier society" can be associated with "technological excellence." If I'm struggling to survive, at the level of getting oxygen
and food, I would be in a somewhat primal state – don't you think? –
that would not at all lend itself to "technological excellence" – which is
absurd! How could anyone think that Martian society could EVER be
based on "technological excellence?" It's just too ridiculous. This isn't
the basis for our existence on Earth – how then could such a thing be
claimed for some theoretical existence on Mars? What a crazy, misguided prediction. This idea of Zubrin's is just another insane confabulation.
But even more shocking! – this seems to be the premise of much of his
ideology!

"Economic links" implies the shipping of cargo. No economic models
could justify the cost or logistics of shipping food, machinery, medical

and scientific equipment, furniture, bedding, clothing/spacesuits, or even *kitchen supplies* to a growing colony. NASA is a charity case and can not pay for these shipments. Or maybe all these things are expected to just pop into existence.

So – Zubrin has defined this rule of thumb: conditions marked by environmental adversity – right to the point – and beyond – of being fatal – lend themselves to "technological excellence." Is this why Silicon Valley is a technological hotbed – because of the hostile and deadly environmental conditions found throughout the valley? Between Antarctica and Silicon Valley, Antarctica clearly has the more adverse environmental conditions, so shouldn't Antarctica – based on Zubrin's rule of thumb – be a veritable nexus of technological excellence? Of course if man insists on living where he doesn't belong, he may "perforce" invent some technologies that allow him to exist, if barely, in that environment – but that certainly does not equate to "technological excellence"...that would be relevant to cultures on OTHER PLANETS. Much less "rejuvenate" – this is hysterical!

Assuming any kind of human civilization could ever exist on Mars, the state of the art on Mars, regarding all things, would at best be at least two years behind that of Earth, at the beginning, and this lag would increase over time, given the increasing self-sufficiency that would be inevitable due to the absurd cost of staying in sync with Earth - as if that was even an assumption. Martians would be living in a different time frame altogether. Of course there would be a need to be inventive, because starting out life on Mars would be very primitive and there would be a desire to replicate Earth's conveniences, but to think that the "state of the art" on Mars could be equal to that of Earth, and that it should remain in synchrony with Earth – this could not happen. Life on Mars would be quite different from life on Earth, each with their unique problems, the solutions also unique. Martian technologies would not be relevant to Earth, and vice versa. Mars wouldn't have the R&D of a third-world country, if any, so movement away from its initial primitive state would be very slow. Martians would be busy enough maintaining oxygen and water supplies, patching their pressurized suits, and other life support systems – things we don't have to worry about on Earth. It therefore doesn't make sense to think that Mars would be a "hotbed of invention," certainly not inventions that would be useful on Earth. It would actually benefit future Martians to be ambivalent about Earth, the two cultures would be that different. Earth would most likely have no economic links to Mars – none. There would be no interplanetary commerce – none. Not with deuterium, not with gold. If Mars was ever to be self-sustaining, which would obviously need to be the case, this means it would necessarily need to be completely independent of Earth. And given its vast, Earth-like resources (according to Zubrin) this should not be a problem!

Zubrin is a man who accepts or rejects ideas solely based on their cost or cost-effectiveness, as evidenced in his book, while completely dismissing the human factor, even ridiculing it – which is ironic because that seems to be the prime objective of the Mars mission – to save humanity.

It's one thing to propose Mars as a refuge of last resort for the sole survivors of some imagined apocalypse, as a planet-sized "bug-out," but to think that it might become a "hotbed of innovation" is simply not realistic – since it's more likely to be the opposite – moreso that Martian inventions would have any use whatsoever on Earth, or that there would be a competition between Mars and Earth to see who could be the most innovative. To think of the creation of a Martian society as a means to trigger an innovation cycle is beyond fantasy. Believing this indicates a tragic lack of understanding of life itself - in *any* sense of the word. If Zubrin and the Mars advocates could only take off their rose-tinted glasses, they would see that Mars is the apocalyptic Earth they're attempting to escape.

13

Interplanetary Commerce

Rockets presume to defy the completely natural principle of gravity, for which it can be said the laws of nature hold us in contempt.

THE UNFEASIBILITY, OR ABSURDITY, OF COMMERCE BETWEEN Mars and Earth requires its own chapter. If there were ever to be a colony on Mars, then due to the complete lack of natural resources of any kind on Mars there would obviously be an enormous need to transport supplies to Mars, that would be proportional to the size of the colony, and so a case could be made to justify this (putting aside for the moment the glaring impossibility of landing heavy cargo ships gently on Mars). But what inconvenience, shortage, adversity, or pain exists on Earth that would be alleviated by "shipments from Mars?" There are obviously none. If Mars had oil, how big would a once every twenty-six month shipment to Earth have to be to make it worthwhile? I'd guess as big as 1,000 supertankers. Is this what you imagine? And that this might be as simple and straightforward as shipping oranges from Florida to New Jersey? If only things were as easy as one imagined. But if a colony on Mars were to be self-sufficient – and we are certainly self-sufficient on Earth, this principle being the basis of our existence! – there would be no shipments of oil from Mars even if such a thing were theoretically possible – because the economics, complexity, and dangers inherent in rocket technology would make it impossible, as demonstrated by the brief history of the Space Shuttle, which was a Girl Scout project compared to what would be involved in interplanetary commerce with Mars. So the shipping of anything *less* useful than oil would be even less feasible – and there is nothing *less useful than oil* on Mars that does not already exist in abundance on Earth – extravagant, rapacious abundance.

Rockets can't be seen as simply another mode of transportation like ships at sea or airplanes. The science of the latter two exploit the laws of nature rather than defy it. Ships at sea exploit the principle of buoyancy rather than defy it. The laws of nature favor our ideas to exploit it in this way. Buoyancy...is free. Airplanes exploit the principle of aerodynamics rather than defy it, and as with ships at sea, the laws of nature favor our ideas to exploit it in this way. Aerodynamics...is free. But rockets presume to defy the completely natural principle of gravity, for which it can be said the laws of nature hold us in contempt – which can not be overcome by clever calculations, noble intentions, or mere mortals entranced by nonstop chants of *we can do it we can do it* – and for which there is an exorbitant and not easily justifiable price to be paid in terms of money, lives, and angst. So a future that relies in any way on rockets flying to Mars seems the most unwise choice of all.

14

The Mars Society

There is no insight to be gained by walking around in a mock spacesuit in the full gravity of Earth, other than the obvious – that they hamper one's mobility and dexterity.

THERE'S AN INTERNATIONAL GROUP OF MARS ADVOCATES CALLED The Mars Society, a non-profit having no government affiliations and established by Robert Zubrin (author of *The Case for Mars*), which has set up some so-called simulated habitats, such as the Mars Desert Research Station, located in Utah since 2002, in the open desert and where participants perform junior-high grade "experiments" using the local geography and weather as the subject matter. If and when man ever gets to Mars, there will – it seems – exist a pervasive duty to perform experiments – which will entail a degree of labor that to us on Earth might appear as a third world sweat shop, or a salt mine – but all in the name of *Science*! To that end, visitors to these habitats are kept busy walking around the vicinity, and riding around on ATVs, in mock spacesuits and playing astronaut – learning "valuable lessons." I'm telling you, I can experience the same "valuable lessons" in my living room after gluing panoramic murals to the walls and walking around in a Halloween costume while talking on my early 1990's cell phone – but that's how I live anyway! Valuable lessons indeed – like learning words like "deployment" and "protocol" – just don't use these words outside the MDRS – please. My residence is more Martian than the MDRS in

Utah even without the murals. The MDRS (or FMARS habitat in north-ern Canada, also operated by The Mars Society) does not in any way, shape, or form, simulate living on Mars. The MDRS in Utah simulates living conditions...in Utah. I doubt if any of the visitors to these "sim-ulated habitats" are in the least bit made aware of the facts about Mars concerning subfreezing temperatures, oxygen, water, radiation, gravity, air pressure, food supply, desolation, et cetera...and might actually be-lieve that time spent in one of these habitats would qualify them to live on Mars! This seems to be nothing more than a failed money making scheme on the part of The Mars Society.

Certainly these "habitats" are nothing but carnival joints.

Zubrin – exposed! Ambitious and industrious, yes, but a carnie nonetheless.

Staying at these habitats to learn about living on Mars is like playing cowboys and Indians to learn about cowboys and Indians – which is ridiculous! All the experiments performed within the confines of these habitats may provide knowledge about the geography and weather where the habitats are located, as redundant as that may be – but noth-ing about living on Mars. There is no insight to be gained by walking around in a mock spacesuit in the full gravity of Earth, other than the obvious – that they hamper one's mobility and dexterity, but this can barely be applied to the reduced gravity and other adversities of Mars. The lessons learned certainly wouldn't be relevant unless the same par-ticipants were the ones going to Mars – which obviously isn't the case based on the projected schedule for a manned mission, still many years away and impossible without several leaps in technology.

The two MARS habitats (in Canada and Utah) – which are nearly identical in design and consist of a 2-story cylindrical building with bare bones accommodations to sleep seven, a shack for power and other util-ities, and a water tank – aren't being used for actual astronaut training – which tells you something – and, despite their exotic locations, are really quite ridiculous. The one on Devon Island in Canada has barely been used since it was completed in 2000, despite the fact that it's fully operational and may provide a more "realistic" experience – which tells you something of the actual interest in Mars – or lack thereof. Is the Mars fever chilling out?

The Mars Society had originally planned for there to be four MARS habitats, but only two were ever completed and put into service. Funding for the program dried up several years before this writing, but funding for space research programs is often tenuous and controversial. Of the two others originally planned, the one for Iceland was built but damaged beyond repair during shipping, another, planned for Australia, was never built due to lack of funds – which means lack of interest. Of the one damaged beyond repair during shipping – could the luck on Mars be any better?

If the MARS habitats simulate Mars (which may include the bonus "full Mars simulation constraints" for a portion of the visit), and The Mars Society argument – which they publicize with utter conviction – is that Mars can support a permanent human colony...then why do the visitors to the habitats only stay a few days, typically between 1 and 3 weeks at the FMARS habitat in Iceland (with 2 groups staying for only 4 and 5 days), and for even shorter durations at the MDRS in Utah? Why can't The Mars Society really prove its point by having these simulated habitats support permanent, self-sustaining colonies – and that grow over time? Isn't this obvious? Wouldn't this be a simple, if not *irresistible*, way to prove their point? Instead – they seem to have proven the opposite! – which of course would be that the MARS habitats (or at least the FMARS) are a failure. (On the whole, any of these MARS habitats might make excellent "bug-out" shelters for use on Earth, or serve as communal living spaces for the homeless.)

Based on the detailed, online documentation and reportage of the FMARS activities (which one can presume to be well maintained and up to date given the unstoppable enthusiasm of The Mars Society and Mars advocates, AND the relative simplicity of maintaining a website), there were NO VISITORS to that facility at all in both 2006 and 2008, and for all other years between 2002 and 2009 there was just one crew that visited each year (with accommodations limited to 6-7 per visit, as with the MDRS in Utah). One can easily conclude that The Mars Society has trouble recruiting visitors to the habitats that offer the most Mars-like experience – which would, of course, include being far away from civilization – such as the FMARS, it being located way up there on an otherwise uninhabited island in northern Canada. But I would expect the opposite – that the most Mars-like habitats would be buzzing year round with campers coming and going! There's a larger experiment going on here, obviously, which suggests that there are actually extremely few people interested in Mars simulations and, by extension, Mars.

It's really all just so silly. Edison is famous for saying that genius is 1% inspiration, 99% perspiration. I'd like to propose that science is at least 20% nonsense.

It seems The Mars Society may have an opportunity they haven't yet seized upon: to convert one of their existing habitats into a full-scale Martian colony – or develop another one for this purpose – that is ongoing and permanent – and self-sustaining – an Earth-based counterpart to what might initially be built on Mars – in anticipation of the manned missions they believe are inevitable – if only to see if a colony of men and woman, accustomed to the riches and pleasures of Earth, then ripped from these – could not just survive but thrive, including raising families, but relying entirely upon such things as hydroponics and other technologies, and ONLY those, that would be implicated for use on Mars, including existing and future earth-technologies that would be carried to Mars...but which would of course afford a lifestyle that would be very basic, if not harsh, even bleak, in comparison, and very labor intensive, without the rewards and long term payoffs to which they would be accustomed.

Because if this can't be done on Earth, it certainly can't be done on Mars. It would at least indicate what kind of people might succeed on Mars – if people with Mars brains even exist.

Such a full-scale colony could, of course, play a vital role in playing out solutions to some of the problems that arise on Mars, that might be more or less transferrable to Mars. Have many of them. These may not surpass the inevitable computer models that will serve the same purpose, but what but a full-scale simulation would be the basis for future computer models? How is it that such full-scale, quasi-realistic simu-lations don't already exist if this is our future and it is inevitable, as the Mars advocates insist. But there's evidently nothing like this currently going on anywhere, just the two existing but underutilized if not derelict MARS habitats, and which seems to be an interesting strategic omission in The Mars Society master plan.

Such a full-scale simulated Martian colony would be stark and not receive supplies other than once every two years and limited to what would be carried on a rocket flying to Mars – as an actual Martian colony might experience. There would therefore be a serious lack of amenities: no pictures on the walls, no cushions to sit on...no alcohol or tobacco (even though people in the full-scale simulated colony wouldn't be em-ployees, these things wouldn't be transported to Mars)...no pets...only things that would actually be transported on a rocket, and you can bet that this would include nothing much besides a food supply that would last at least two years and that would be in various states of dehydration, basic personal hygiene supplies, nutritional supplements including vitamins, artificial flavorings, medical equipment, antidepressants and mood elevators, physical fitness equipment comprised largely of those stretching elastic bands, equipment to grow plants hydroponically and the tools needed to assemble these (they will, of course, be self-as-sembling to some degree), and some larger tools but manual because electrically powered tools will break or malfunction. It would be a bad idea to have the colony acclimated to sophisticated (modern) tools and technology because this complexity adds the need for service and re-pair, which implicates several more dimensions of infrastructure that is probably beyond the scope of at least the first generation or several of man on Mars. And of course the residents of my proposed, full-scale simulated Martian colony would be permitted – encouraged! – to have and raise children. Because that's what life is all about, whether it's on Earth...or Mars.

But I can hardly imagine such a thing – a simulated Mars colony on Earth, with people living – voluntarily! – this way. So it's even harder for me to imagine people living on Mars – going mad but for their spinning dial-a-flavor!

THE HABITATS ARE A FRAUD

As it is, these so-called habitats are not simulations, and it's irresponsible for The Mars Society to make that claim, which will do nothing but mislead not only the habitat visitors but everyone proselytized by The Mars Society. This is another example of the brainwashed mentality of the Mars advocates. Nothing has been learned in these MARS habitats that can be taken to Mars or is relevant to living on Mars. To put another way, a conventional budget motel at any of the two MARS sites would be just as educational – or spending any extended period of time under house arrest in a penthouse suite! Because these MARS habitats have all the amenities of a budget motel, and none of the fatal adversities of Mars, nor even the lesser adversities – nor simulations thereof. Even where some adversity could be simulated, such as with the food supply – there is the opposite. Not only do all these habitats have a surplus food supply, but it's regular food, and microwavable – even on the ISS (the manned space station in orbit around the Earth) they don't have a microwave oven! There is not a single aspect of these simulations that simulates Mars! The MARS habitats are all really quite fraudulent. The habitats are not simulations – they don't simulate, nor even attempt to, any conditions known to exist on Mars or that would be experienced by people living on Mars.

Any simulation that occurs is entirely – 100% – in the imagination of the participants.

Any of the two MARS habitats may as well be exhibits at a cheap amusement park, and even for that purpose there's nothing to be experienced! These two habitats may simulate what housing conditions could be like in some future, overcrowded Earth, but as a simulation of living on Mars – a complete fraud. It's a simulation that might be set up by a bunch of high-school kids based on a 1950's sketch – before we actually knew anything about Mars – certainly not what one would expect from an ex-NASA engineer. Even bad science sometimes has a certain intelligence or quirky charm to it, but the two MARS habitats are just plain dumb. And if The Mars Society can't get it right – with their prestigious and high-tech connections – who can?

Regarding pictures on the walls: this and other examples of art and culture may seem trivial, but the human spirit, besides evidently the need to soar into space, needs occasional glimpses of art and beauty – and sometimes even to create such things – otherwise we're just animals, not human. Humans like pretty things, we like to decorate – to liven up! – our living spaces, and the world around us – and so the means must be devised to fulfill this aesthetic. Even cavemen had drawings. So, do we include the means to satisfy this aspect of our humanity in the payload of rockets bound for Mars – or take the risk that humans on Mars may become soulless automatons?

Biosphere 2 was another so-called self-sustaining ecological experiment, built in rural Arizona and which ran from 1991 to 1994. It was comprised of several futuristic looking buildings built for the purpose that enclosed 3.14 acres, had living quarters for 8 people, and was a highly sophisticated ecosystem – even the air and oxygen supply were to be self-sustaining. Residents of Biosphere 2 were not only not disadvantaged in any way (such as through a lack of food, social isolation, inactivity, or circumstances meant to test endurance or adaptability) – they had every advantage of modern man! – in beautiful, sunny Arizona! This "self-contained" ecosystem might even have *enhanced* the experience of being human – such that the residents would refuse to ever leave! *Instead, the opposite happened.* Analogies to a Martian colony are obvious, but where there will be none of the benefits of modern man, nor even air or water. Interestingly enough, Biosphere 2 suffered from carbon dioxide levels that fluctuated wildly, along with diminishing levels of oxygen, and most of the vertebrate species and all of the pollinating insects died, according to Wikipedia. Could things be better on Mars?

The purpose of these, one might say – grand – experiments is to see if we can separate ourselves from the world, while still enjoying healthy and fulfilling lives, but they all fail to some degree or another, sometimes disastrously. It is axiomatic that Earth is self-sustaining, but it stands to reason that if you isolate or separate it into parts, that these parts will deteriorate and eventually die, and these experiments seem to bear this out but they are sometimes interrupted early on, at the first signs of trouble – but not so late that they might still be considered a success. These experiments effectively prove how complicated the relationship is between living things and their environment, and that variables that one might think would either cancel out or not be a problem sometimes end up being just the opposite – important matters concerning the food supply, whether or not it provides sufficient calories and variety – and other, simpler issues, from the technical to the administrative.

When Biosphere 2 ended prematurely for such reasons, the participants just walked away. If a colony on Mars doesn't go as well as planned – and how could it? – the participants would not be able to just walk away.

It's interesting to note that Hollywood treated the Biosphere 2 concept as a parody, in the movie *Bio-Dome*, portraying it as a goofy comedy. One might wonder if there is really any other way to look at it.

15

Antarctica is More Like
Malibu than Mars

It seems there'd be a thriving city of a million people – at least – living in Antarctica before 80,000 were living on Mars.

THERE AREN'T ANY TOWNS IN ANTARCTICA, BUT HUMANS WITH A frontier spirit have established a few base camps, for researchers who have forsaken the sunny warm climate ideal for one that is extremely cold and frostbitten and who rely exclusively upon canned and prepackaged foods, with no deliveries during the winter. Only a few people have ever been born in Antarctica.

The Mars advocates have convinced themselves that Antarctica is like Mars, that Antarctica is the closest anyone on Earth can come to living on the surface of Mars. But let's be clear: even if this were true, relatively speaking, it doesn't mean it's really anything like Mars – because Antarctica is nothing like Mars. Antarctica's permanent human population is zero and it's considered to be uninhabitable, mainly because of the desert dry air (getting very little snow or rain, the icepack being the accumulation of eons) and extremely cold temperatures (41-59 degrees F maximum in summer near the coast, with record low winter temperatures of minus 112 to minus 130 degrees F toward the interior), with 24/7 sun in the summer, 24/7 darkness in the winter.

Don't tell this to the Mars advocates! Draw a line, with Mars at one end, Malibu on the other. Imagine the line is a mile long. Yes, that's Malibu, California, world famous for its sand, surf, and sunshine (and not the other Malibu). Now, where would you place Antarctica along this line, based on how closely it resembles either Mars or Malibu? If you're a Mars advocate, you'd put Antarctica somewhere near Mars, thinking they're both desolate and freezing. But it's not all that obvious, if you've been brainwashed by NASA and the popular press articles written by "science writers." Actually, Antarctica would be about a foot from Malibu on our mile-long line, with Mars way at the opposite end! **Are you in shock?** That's how closely Mars resembles Antarctica – it doesn't! Because there are other things besides temperature you have to consider when you're comparing Earth and Mars! – and the Mars advocate space cadets choose to ignore these other things, the *unearthly* values of all the other relevant variables and circumstances on Mars, all of which have been reliably measured and verified to the nth degree of precision. By NASA. The extent to which your placement of Antarctica is further than an inch from Malibu on your line demonstrates not only the confusion that exists regarding the potential of Mars as a habitable place, but the effectiveness of the Mars advocate space cadet brainwashing propaganda.

Put another way, on a 12-inch scale representing Earth, as a measure of livability using normal standards, Malibu and Antarctica would be at opposite ends – obviously. In the normal sense, they're opposites. If you wanted to put Mars on this same scale, how far out would you have to extend this scale on the Antarctica side? At least a mile – obviously! – *but not to a Mars advocate.*

Here's yet *another* way to look at it: Man and monkey have a DNA equivalence of about 90% – and you see how different the two species are. By way of analogy, Mars and Antarctica have a DNA equivalence of perhaps 1%. Even if we increase this to 50%, one can hardly justify the conclusion – or premise – that "what's possible in Antarctica is possible on Mars," or similarly misguided logic.

Antarctica is a veritable paradise compared to Mars, with normal amounts of oxygen and water. To anyone on Mars, Antarctica would be Heaven! And to anyone in Antarctica, Mars would be ...Hell!

Even if one were to *accept* the comparisons of Mars to Antarctica, it would have to be on the condition that Mars has something that Antarctica lacks – which is certainly not the case, based on all the evidence, in all its degrees of precision. And given that "almost nothing" grows on Antarctica, anywhere, it stands to reason it could only be worse on Mars, in view of the subfreezing temperatures alone – even if it may get "above freezing" during the day during some fractions of the year since the year round night time temperatures alone are well below freezing. The "almost nothing" refers to a variety of highly adapted members of some lower plant groups such as mosses, lichens, and fungi found on the 1% of Antarctica, along some parts of the Antarctic coastline, favorable enough just for that (http://www.antarctica.ac.uk/index.php), where the climate is generally warmer and wetter – which is an oxymoron on Mars and therefore even this kind of life isn't possible even if cultivated.

However, the population growth of Antarctica since its discovery may be an excellent reference point in predicting the future population of Mars – they are at least both frozen deserts. The fact that Antarctica has stopped growing in population, which is the coldest place on the planet where people can live, suggests that there is a lower limit to just how cold things can get and remain livable – because there isn't another reason its population has leveled off.

The population of Antarctica is about 5,000 – during the summer, and consists exclusively of scientists. But it drops to about 1,000 during the winter, when it's much colder (like minus 94 F), then it returns to about 5,000 the next summer.

The population drops in the winter *because* of the winter cold – it's not just a coincidence. But these are scientists – on a mission! If they can't tolerate temps in Antarctica as cold as an average of minus 94 F in winter (comparable to predawn temps on Mars) for just a few months, how can they tolerate these temps on Mars where they're year round – and where there is also no oxygen or water? Do the scientists who go to Antarctica in the summer but leave in the winter have an attitude problem? Why would they leave just because it gets colder? OK, let's give them a break – and conclude instead that the winter temps in Antarctica might actually be outside man's capacity to acclimate (although this is not the same as genetic adaptation but which takes generations) – so it's hard to maintain any sense of optimism for success on Mars regarding this particular variable, it's just too cold. But this is an understatement, because you couldn't ask for a more convenient – and obvious – reference point.

Indeed, if they can't figure out how to live comfortably in the comparatively balmy Antarctic winters – which might include wearing spacesuits – as shown by the fact that most of the scientists vacate during the winter months...how could an even greater population live comfortably on Mars where it is much colder and where there is no oxygen or liquid water?!

Sure they'll have spacesuits on Mars but they can wear spacesuits in Antarctica! So that cancels out the spacesuit counterargument. Either way, the max population of Antarctica in wintertime, which seems to have reached a critical point, is obviously valuable in theorizing an upper limit to Mars' future human population – while being aware that – according to scientists – Mars is getting *even colder...*

PROXIMITY ARGUMENT

Ignoring all the other differences for the moment between Mars and Antarctica, one big difference would always remain: proximity. Antarctica would have the advantage if only because it's closer.

No matter how many people lived on Mars, it stands to reason there would be that many more in Antarctica

and proportional to the convenience afforded by proximity alone (even if there was no oxygen or food in Antarctica) since neither has an indigenous human population. Most of us regard Mars and Antarctica with a strong indifference, but if I had to go to either it would of course be the closer of the two. So Antarctica would always have a larger population than Mars – it would always be easier to get to.

Now, given the relative convenience of Antarctica, one might have expected a virtual explosion of growth – certainly once the basic foundations were established in the mid 1900's. But time has told us that this is most definitely not the case. Because the population of Antarctica, rather than growing at a steady rate, reliably waxes and wanes as described in the previous paragraph, which has been the general pattern for the years leading up to about 2012 – and no surge is planned or expected. (What? Can't we terraform Antarctica so it's livable year round?) And there is no trade between Antarctica and other countries – despite the relative ease and "potential for rejuvenation" that might exist in the same "visionary" sense that some have proposed for a Martian colony (specifically, Robert Zubrin in *The Case for Mars*).

So, based on all this, in consideration only of the enormous disadvantage created by our distance from Mars, what might be the potential MAX population of Mars, using a simple extrapolation? Even in a best case scenario, significantly LESS than the 1,000 population of Antarctica during its coldest months, when it is more like Mars, obviously – or about 50. Please insert your guess if you think it's better. But Antarctica and Mars have many more differences than their proximity to us – and they're all on the negative side for Mars! And it turns out the cold temperatures, as dangerous as they are, may be the least of the Mars advocates worries. Poor Mars. So this number would decrease further still from 50 – enormously! – when you figure in only the most obvious of the other negative factors on Mars like no oxygen or liquid water or food. To barely enough to comprise *any* size colony! To maybe just enough to send up to *scout around* for a while. In this light, considering that the rovers are probably *better* at scouting around than man could ever be, *there would be no point for man to go to Mars at all.*

One might, then, consider this – unmanned – exploration to be the pinnacle. Unless one is hell-bent on seeing for oneself just how badly suited Mars is for habitation.

Put another way, it seems there'd be a thriving city of a million people – at least – living in Antarctica before 80,000 were living on Mars, since it seems quite plain that the population of a place is proportional to its livability and accessibility. Do the Mars advocates think Mars is more livable or accessible than Antarctica? As you can see, Antarctica has many things going for it, compared to Mars. Antarctica is relatively easy to get to and there are no government restrictions on its population growth. Supplies can be relatively easily shipped in. Yet Antarctica is considered to be unlivable! Why would this be the case? It seems that, based on a scale of livability, ANY town on Earth with a population of less than 80,000 – would grow to 80,000 – and beyond – before there would be a colony of 80,000 on Mars. Suffice to say, in this regard, that a colony on Mars would have much fewer residents than even the smallest town on Earth.

And when scientists (and other workers) temporarily move to Antarctica, there's no pretense that it's for another reason besides scientific study or that they might be staying on as permanent residents. There's no confusion about its unlivability as a permanent residence, everyone stands in agreement about this – there are no holdbacks who refuse to leave at the time of their planned departure. In other words, the scientific community obviously understands, and appreciates, that there are limits, beyond our control, that determine...whether or not we can survive in any particular environment – it's not just a matter of will power or determination or engineering chutzpah. Which leads to the same conclusion (as the proximity argument), that if the max population of Antarctica when it is most like Mars is about 1,000, then the max population of Mars would be significantly LESS than that, and far short of the 80,000 or more envisioned by SpaceX's Mr. Musk.

The most interesting thing about Antarctica – as it pertains to the case I'm making, of course, because there's actually nothing interesting about Antarctica – is that anyone would want to live or work there at all. It's a resume builder, for those collecting merit badges. At the political level, all the major countries have flags flapping in the Antarctic wind – let's not let any one country get too imperious or take the idea of world domination too seriously.

COULD THE MARS ADVOCATES SATISFY THEIR PENT UP MARS MANIA by moving instead to Antarctica? Think of it as a *Mars substitute*. Antarctica is luxurious compared to Mars. This could be a short term solution to their bubbling...their seething mania...for Mars...their discontent with being on Earth. What punishment, to be an Earthling. God (or whatever intelligent designer) gave us our own planet but *they don't want it* (which works into the Mr. Musk as antichrist idea quite well). Antarctica is freezing cold, desolate and lonely – which seems to be just what they're looking for, plus a few other *amenities* like oxygen and water and air pressure, which precludes the necessity of a pressurized suit – we need to convince the Mars advocates that this is on the plus side, in addition to the isolation that causes part-time residents of the base camps to claim feelings of being "dead" and "not real."

A paper cut on Earth would not be a paper cut on Mars.

Planes already land at the various Antarctic base camps during the Antarctic summers, so anyone can go to Antarctica without years of couch potato killer training, which is obviously convenient, if you're a Mars advocate with a bubbling, pent up mania to live on Mars. Another advantage to using Antarctica as a Mars substitute is that supplies can be flown in – and a life that retains complete synchrony with other Earthlings – without the need for technology that is yet unimagined if not impossible as would be the case in living on Mars. Antarctic residents can call friends and family back home and speak to them in real time, unlike on Mars where there would be a lag time of 6-42 minutes that makes talking on the phone (to Earth or people on other planets) impossible. One advantage of real-time phone conversations to people in Antarctica? Your doctor can talk you through an operation – that you're performing on yourself. This happened in 1998 when a women (who was herself a doctor) with breast cancer in Antarctica operated on herself. And another time in 1961 when a man, also a doctor, had to remove his appendix. People on Mars would not have this capability – to call a doctor on Earth – would never be able to call Earth for any reason whatsoever. Could never have supplies flown in or delivered in any other way in an emergency, no matter how dire the circumstances.

"The door slammed hard right on his face –
giving him a concussion."

This could have been headline making news, in August of 2012, but wasn't.

Antarctica's Concordia base station normally experiences temperatures of minus 112 degrees F and wind chills of minus 148 degrees F. Of the twelve people stationed there from February through November, 2012, one was a doctor, Alexander Kumar. Due to his slowed reflexes – by his own admission – as occur in Antarctic winters, a door, whipped by the wind, slammed into his face – and gave him a concussion. He was "down" for three days. A concussion? The doctor? THE doctor? I won't say "I told you so" regarding anything that can go wrong, will go wrong. He was HOPING it was just a concussion! Dr. Kumar himself later suggested there should be two doctors at the base. Just two? How about every single person stationed in such remote confines has extensive medical training, including training in basic surgical techniques. Do you see how quickly, how SIMPLY, things can go terribly – catastrophically – wrong? And if things that wouldn't likely happen anywhere else (according to Dr. Kumar), happen in Antarctica – and entirely due to its "unearthly" conditions! – this tendency towards unearthly problems and mishaps *could only be magnified on Mars.* It's only too easy to imagine the same thing happening on Mars, with a gust of wind causing a door – or something else – to smack into the visor of a spacesuit helmet – where the slightest crack could lead to an immediate and rapid depressurization – and death.

- Dr. Kumar stated "It's almost as if our senses become understimulated and wither in the darkness, ice and silence. So when a new stimuli comes along it can be disproportionately fascinating.

- He thought about leaving Antarctica *from the first day.*

- "We have one female crewmember among a crew of 12 men. There is the potential for it to cause problems – previous Antarctic missions have been plagued by jealousy." [http://www.smartplanet.com/blog/cities/preparing-for-human-mars-mission-the-loneliest-job-in-the-world/4589]

BLACK SWAN EVENTS

This door-slamming-in-face accident didn't involve testosterone fueled hotshot behavior, bad judgment, improper or inadequate training, or the other usual culprits. It could be considered a "black swan" event, according to the probability theory developed by Nassim Taleb, which deals with highly improbable events. It would be a black swan event if the same thing occurred on Mars and lead to the demise of the colony, which seems "highly improbable" to those of us on Earth but things will be more fragile and precarious on Mars. Whether or not something is a black swan event all depends on the point of view of the observer. The infamous space shuttle disaster caused by failure of the O-Rings could also be considered a black swan event because it was highly unexpected, highly consequential, and easily rationalized in retrospect – these are the defining characteristics of a so-called black swan event. That's why Neil Armstrong was only on the Moon for two hours – they didn't want any black swan events screwing up the mission! We simply can't anticipate or predict every eventuality – regarding events that are, or considered to be, completely outside the realm of possibility, and we can't possibly begin to comprehend the scary or improbable eventualities that absolutely and without a doubt exist on Mars. But if you do attempt to comprehend this, don't forget to factor in hotshot behavior, bad judgment, improper or inadequate training...and the other usual culprits.

Imagine how easy it would be to sabotage a colony. All you'd need is one rogue human-Martian gone berserk – or simply having a bad day, his metabolism not adapting, a bad batch of air or grub...and that's it.

16

The 78,000

This kind of knee-jerk reaction to a world-wide ad of this kind is not a surprise – in which is released the pent up mania...of centuries!

IT'S INTERESTING TO NOTE THAT IN APRIL **2013**, IN A TWO WEEK period, 78,000 people expressed interest in going to Mars, one way – never to return – in response to an unprecedented online ad for "inhabitants wanted" placed by Mars One, the Dutch "concept" previously mentioned. At first glance that might seem impressive, and contradict my pessimistic outlook, but it's easy to challenge the seriousness of these 78,000 (or however many hundreds of thousands may have eventually responded to the ad during its 4 month run through August 31, 2013).

FIRST: of course it seems exciting – the idea of visiting another planet! Never before in the history of mankind and all that. **Freak out!** This was a truly historical event – as singular as the Big Bang itself and never to be repeated! – one that would generate the interest of school students, or at least science classes, worldwide. Imagine entire classrooms full of excited students signing up en masse – what a gas! It's impulsive fun to respond to this kind of thing – while at the same time without making any real investment or commitment – while still avoiding the reality of abandoning one's family (and to never even speak to them again!), favorite foods (the basic concept of food would be completely redefined), all their favorite things, and their beloved pet. Many young people have vague or unbridled ambition and no sense of danger, while others are unemployed and see it as a "wonderful opportunity." Others are just curious, or intrigued by the prospect of being "selected for astronaut training" – in the same way I once applied to NASA to be an astronaut. It's fun just applying! Applying to be a Mars astronaut has great potential as a fad, a craze, where it becomes a status symbol just saying you applied. Who knows, it may become a rite of passage. I may apply, just to see how far I can get. Wouldn't it be ironic if I was chosen for training? No no, I was just kidding! And this would all be a *problem* for Mars One, because many who will apply are not serious and this will bog down the recruiting process enormously. Especially if people expect to be paid during training, and recruits treat it as a regular job, then back out at the last minute – don't think it won't happen. How will Mars One know who is truly committed? Perhaps recruits should not be paid during training. Or recruits may have to pay some huge amount up front, besides the application fee. **Because it's crazy to get on a rocket bound for Mars knowing you'll never return and what insane fool would do that?**

Second: almost all the applicants think that Mars is habitable because that's what the propaganda says (like the "Colonization of Mars" Wikipedia page – which has global readership – that says Mars is "hospitable," and about which I say more elsewhere in this book). These initial applicants obviously have no idea that Mars is far less habitable than even Antarctica – a place deemed uninhabitable, and a place they would never visit – could never visit.

Third: the typical applicant believes they're signing up for the round-trip excursion package with all the touristy frills. They think that a trip to Mars is not just technically possible – which it isn't – but even that it may be "safe and ordinary." Well finally! And of course Mars One would not put them in harms way! Most who sign up don't realize that they will be, literally, test pilots in completely uncharted and terrifying territory – and will most likely die at some point in the process, perhaps just in the training.

Fourth: the application process was via the internet, which only makes it too easy for those who aren't really qualified, or just curious, to apply.

FIFTH: I'll just go ahead and group the other reasons together: Those Arizona-esque NASA photos of Mars are so adorable, with their warm glow – who would not fall for them? Some who sign up probably think they're signing up for a "simulation" of a Martian trip. They don't understand that just getting there will take 8 months – all of which will be spent in sheer, unrelieved boredom – of a kind that is literally not to be found anywhere on Earth. They all assume that there will be a minimal set of creature comforts and ample provisions, during both the flight and on Mars – that they will be provided for and perhaps even pampered. They don't realize they will have to give up all their favorite things forever. Lastly, they all obviously know nothing about Mars. It is, after all, being promoted as FUN FUN FUN even though it is NOT NOT NOT and will in fact be quite the opposite! We may as well just go ahead and disqualify every single applicant because by now we've covered all the bases.

So this kind of knee-jerk reaction to a world-wide ad of this kind is not a surprise – in which is released the pent up mania of *centuries*! – it only confirms that there are a lot of Mars advocate space cadets. But this, of course, does NOT equate to a successful Martian colony, nor the prospects for one, and given the size of the application fee, which varied by the wealth of the applicants home country from $5 to $75, it reeks, at least a little, of a money making scheme – otherwise I'm buying stock in Mars One and hope that millions more sign up! Anyway, according to the ad, those selected would require eight years of training, with a planned launch – to Mars – in 2023.

If all the applicants to the Mars One project were told that they had to start working out two hours a day as soon as they applied, there would be about 95% or more *fewer* applicants, assuming that only those who already had a regular exercise routine would apply – unless you are a Mars advocate and believe in the impossible: that any of the others would just get up and start working out *two hours a day*. How's that for bizarre fantasy! Of the remaining 5%, presumably all gym rats, they should enjoy working out not for the usual benefits but just to stay alive, be able to take on the role of doctor, and enjoy – *really enjoy* – laying bricks. The point I'm making, of course, is that *there is no such person!*

BUT WAIT! OF THE 78,000 FIRST REPORTED, ONLY 7,000 COMplete applications were received by Mars One – which would include the application fee – but of course. You can't imagine how learning this satisfied the pessimist in me. Who does anybody think they're kidding? And with each step further along in the recruiting and training process, this number will drop *exponentially* – because only in science-fiction could there be a colony of humans – a civilization – on Mars. But let's keep ignoring the realities for the moment, and continue to imagine.

When all the people who signed up with Mars One for the trip to Mars learn about and appreciate the Round-Trip Fallacy – which means they would never see their spouses or families or friends or pets or favorite things again (and which I explain in another chapter) – they will most likely become Earth advocates. This would cause a significant number of dropouts from the Mars One training program – and could potentially be the deal breaker for the entire program. So Mars One will want to keep this knowledge from their recruits – which is of course a form of propaganda: keeping the truth from people. But perhaps it will be the friends and families of the recruits who intervene – it's just so sociopathic for a normal person to just give up all these things – or to want to leave Earth *with little or no prospects of returning!* – it's just so unnatural and *repugnant*. Every single applicant to Mars One will eventually confront this reality, now that the *pent up mania* has been released, and only a true deviant or sociopath would be unaffected by it. That's one of the dirty little secrets of Mars One and the other Mars programs – now fully exposed.

It's one thing to be interested in space generally, and it's a great spectator sport. Many people will accuse me of being overly critical, or a pessimist, but you can't extrapolate the success of a Martian colony from the enthusiasm for the numerous lunar missions or the impulsive response to an ad by Mars One. Getting to the Moon was easy – it's much closer, there was never the intention of staying there,

and the astronauts were in constant communication with Earth – which is perhaps one of the most unappreciated aspects of the lunar missions, the ISS, and Space Shuttle projects – but which will not be possible with missions to Mars or beyond.

At the least, only after some kind of suspended animation is perfected would the trip itself be feasible, since it would take eight months. Who wants to sacrifice eight months of their life just for travel time, and that's just one way. According to NASA the trip would, in fact, be one way, barring as yet unforeseen breakthroughs that would be on a par with the Wright brothers or Edison's ingenuity – or Einstein. There's no established way for people to land safely on Mars (heavy payloads are impossible to land gently as of this writing) and there's no technology that would allow for a return trip (among the endless list of other things) – all of which was known by Mars One, but which didn't stop Mars One from the first running of their provocative ad for "inhabitants wanted" in April 2013.

17

Mars One Training

There could be any one of a hundred or more different scenarios playing out upon touch-down, but not one from the dozens they had expected and rehearsed.

So, ACCORDING TO MARS ONE, THERE WILL BE EIGHT — 8 — YEARS of training. This seems like a long time, *especially if Mars is already habitable*! – as the Mars advocates claim. A lot can happen during eight years, including many life-changing events. Will I be allowed to have a normal life during this time? This would include (for those not already married) dating, getting married, buying a house, and having children. And who would marry someone if they knew they'd be leaving for Mars in a few years, never to return – would NASA provide child support? Would I be allowed to do all this on Earth if I was expected to have another family once on Mars? This would be a pretty weird conflict of interest. Would Mars recognize a marriage on Earth, or would I need to get divorced first? Can I be married to people on different planets? This would, of course, take the idea of "down low" to the next level. What about my credit cards and other debts? A person could live a pretty indulgent, credit-based life if they knew it would all be erased as soon as their rocket launched.

And who will train these astronauts? There's a certain paradox in this, because no one's ever been to Mars! And there is really nothing on Earth that is equivalent to living on Mars – even living on the International Space Station (ISS) is hardly relevant to living on Mars.

"How on earth" could such a curriculum be put together? The difficulty in defining it can't be overstated. Just imagine: "Now remember class, the first thing to do when you step down onto the surface is..." There could be any one of a hundred or more different scenarios playing out upon touch-down, but not one from the dozens they had expected and rehearsed.

How could anyone keep eight years of training in their head? This couldn't be eight years of boot camp-style training (which, for entry into the US Army, is only nine but intensive weeks) and would require something far more leisurely. It will (or should) probably include two years of real medical school – which bears a special irony because most in this training probably wouldn't otherwise be qualified to get into medical school, nor would they have the interest, so having a medical degree or other specialized training already would be an advantage.

But it seems that the training program would not actually last eight years, that there would be sufficient candidates that already have the equivalent intelligence and skills

because that's a lot of overhead, to train every single trainee for eight years, during which you'd also have to expect an appreciable number to drop out. Over the course of eight years, I would expect the drop-out rate to be very high, especially for those who wake up to the fact that Earth isn't such a bad place after all.

It might make more sense, be more expedient, to provide some "basic" training, have them read this book – which would all take less than a year – then send them up and see what sticks.

Is it possible to sufficiently train a crew of 3-4 astronauts (the size proposed by Mars One) such that they can function entirely independently of a ground crew? This is what would be required of a mission to Mars – to function independently of a ground crew. All the lunar missions were in constant, real-time contact with a ground base of literally hundreds of specialists that could offer their immediate assistance or knowledge – this was critical for the lunar missions to succeed and is easily taken for granted.

WHAT IS THE MINIMAL SIZE CREW THAT COULD HAVE THE KNOWL-
edge and competence to handle literally any unexpected adversity, catastrophe, or inconvenience *just for the trip*? There may not be one. The amount of training for just a few would be mind-boggling. Or you could have a staff and crew of a hundred specialists, travelling together, to guarantee a reasonable expectation of success, but is "reasonable expectation of success" enough? A switch is short-circuited, there's a leak in a canister, something freezes up, a sensor that is supposed to detect a malfunction doesn't work, a camera stops working, a manual override doesn't work or doesn't work as expected, some of the food supply has gone bad – the list of things that can go wrong and that would need to be fixed is...incomprehensible. The slightest mishap or foul up could spell disaster and instant death to the mission. Neither the lunar missions of the 60's and 70's nor the subsequent shuttle missions have demonstrated that we can pull off such a feat – functioning in space without a dedicated support staff on the ground. So what makes us think we can do it now? Could we build a rocket to carry a hundred men and women safely and reliably to Mars? Then expect them to live there indefinitely? And how do you suppress the urge to hurry up and get back home, as was the case with the lunar landings, the shuttle trips, and the ISS astronauts?

18

Human Nature

Our continued fascination with Mars demonstrates, if nothing else, how we hang onto both the past, and to antiquated visions of the future. That's right – our vision of man on Mars is not futuristic, but antiquated.

THERE ARE FOUR CENTRAL ASPECTS OF HUMAN NATURE THAT preclude us from living on Mars, which I'll summarize here and elaborate upon throughout the rest of the book. These are somewhat separate from, and may considered transcendent to, the issues regarding the rocket science (which is merely a distraction) and the deadly environment of Mars.

1 - Biology

We are really nothing but highly complicated biological organisms – who are also self-conscious and self-directed but these are incidental to our biology – which inextricably ties us, as with any living thing, to a very specific ecosystem.

The Mars advocates have this reversed, thinking we are primarily willful creatures whose biology is incidental.

Only after our biological needs are satisfied can our willfulness be satisfied, and Mars is completely unable to satisfy the prerogatives imposed by our biology.

2 - Fragility

The fragility of both the human body and mind, our preoccupation with longevity and quality of life, and the unavoidable priorities this causes to determine our existence on Earth. We don't and can't control the elements of our ecosystem – through will power or any other means, otherwise we'd turn the world into one big giant Malibu – although we may have done the equivalent with our artificial, Earth-proof environments – our homes. But this only demonstrates how finely tuned our biology is. We could never do this literally – we haven't, we aren't, and have no plans to do this – because such a thing would be preposterous and insane.

3 - Materialism

Life itself is an anomaly, the odds against our existing at all are incomprehensible, but somehow life has sprung forth, and manages to persist – but through a form of compromise, in which we, our bodies, are in a constant state of deterioration; nature fully intends and conspires to reclaim us, and man is in a constant struggle to combat this deterioration. We've invented many ways to do this, namely, the machinery of our lives. *But none of this machinery will exist on Mars!* This means that the natural deterioration we experience on Earth would, on Mars, only be worse, in a steady state of FREE FALL – unabated – unstoppable. Plus we love *stuff*. There's no *stuff* on Mars!

4 - Creativity and the need for fulfillment

Mars is boring. Mars is...*Desolation*. Any thrill would be gone within days of getting there, probably sooner, at some time during the eight month trip there. The human spirit, besides evidently the need to soar into space, needs occasional glimpses of art and beauty – and sometimes even to create such things – otherwise we're just animals, not human. Humans like pretty things, we like to decorate – to liven up! – our living spaces, and the world around us – and so the means must be devised to fulfill this aesthetic. Even cavemen had drawings. So, do we include the means to satisfy this aspect of our humanity in the payload of rockets bound for Mars – or take the risk that humans on Mars may become soulless automatons?

19

Bodily Deterioration

Actually, we didn't adapt to Earth, that would imply we're somehow separate from it. It's more accurate to say we're a manifestation of the laws of nature (as they exist on Earth), rather than something that merely adapted to the laws of nature. Thus we have no choice but to survive, and thrive – as earthlings we have been sentenced TO BE. But Martian laws of nature are different from Earth laws of nature – the physical constants are different, like gravity. We already know that Martian laws of nature are not conducive to human life, in fact are extremely antagonistic to human life and possibly all life no matter how simple. Regarding a human existence on Mars, we have been sentenced to NOT BE.

THE HUMAN BODY REACTS PRETTY BADLY TO SPACE TRAVEL — BEcause of the reduced gravity in space – and deteriorates, in some ways that are irreversible. Reduced gravity causes a loss of muscle and bone mass proportional to the reduction in gravity. The astronauts are granted no mercy from this – no amount of training or willpower can defeat this – and this will be true for anyone who travels in space, whether to the Moon, Mars, or anywhere. This is also true if you're simply orbiting around a planet, like scientists living in the ISS, which orbits the Earth. The longer you're in space, even if just floating in the International Space Station (ISS), which has been orbiting the Earth since 2000 at a distance of only 250 or so miles above us, the more the body deteriorates, which of course causes all kinds of problems both immediately and when you get back to Earth. Like not being able to stand. Astronauts actually have to be carried out of their landing vehicles when they return to Earth's gravity. We usually don't see this on TV or in photographs, so there is a little bit of an illusion concerning the whole thing – one might say propaganda. NASA has a way of keeping certain unattractive things on the hush-hush. But doesn't that violate the objectivity of science? – in emphasizing certain data while concealing other data? That seems quite unscientific. Astronauts know this – that muscle atrophies in low gravity – even before they start their training, it's not as if NASA is withholding this bad news from recruits – right?

Fortunately, regular daily exercise, about two hours a day running on a treadmill or standing in a hyperbaric chamber and other exercises, can compensate for some of the bone and muscle loss. Regular exercise plays an enormous part in the life of astronauts on the ISS, and this will be true for anyone living on Mars. Whatever other negative effects of the reduced gravity on Mars are yet to be discovered, and whatever forms of compensation that can be devised, if possible, will be part of the adventure.

Anyway, just in the trip to Mars, bone demineralization will be significant, besides the muscle atrophy and, to a lesser degree, vision loss – another consequence of space travel that NASA isn't too public about. For astronauts returning to Earth, normal physiology tends to return over time, but bone recovery has proven problematic. For a three to six month space flight, it might require two to three years to regain lost bone, but in some studies, bone loss was not recovered.

20

You Like to Eat,
Don't You?

Let's not understate the moment the first seed might germinate on Mars, as unlikely as this is. And so might begin the first page of the first chapter of the first book in the first library of a new world.

WE KNOW THAT THERE IS NOTHING GROWING ON MARS — NOTH-ing whatsoever. Not even at the microscopic level. And we know that we can't just throw down some seeds and expect them to grow, even if cultivated, since there's no water in the ground in which would be dissolved the vital minerals necessary for plants to grow, and which would comprise the bulk of the plant material – after all, plants, and all living things that we know of, are mostly water, about 75% for plants but some as high as 95%. A plant that could extract and store large amounts of water from the atmosphere or growth medium could be a primary source of man's daily water requirements – but far be it from Mars to be trusted to provide us with such a convenience.

One thing is for sure: we'll need to grow plants on Mars if we want to live there. This is the biggest challenge facing man on Mars, in parallel with the freezing cold, water, oxygen, and other critical issues, and is certainly an evocative one – to think that Mars could come alive! Which is what would happen if things were sprouting from the surface of Mars, or even from plastic containers or troughs, row upon row of them, housed in shelters below or on the surface.

Any life on Mars would be anaerobic – not require oxygen. To that end, are scientists developing anaerobic plants for future Martians to eat? Unfortunately, there is actually no such thing as "anaerobic plants" (true anaerobes only exist among bacteria and fungi) – which makes sense since, given the abundance of oxygen on Earth, "anaerobic plant" would be an oxymoron! Even on the microscopic level there is nothing promising in this area due in some part to the hugely unappetizing nature of anaerobic bacteria (which cause several human diseases including salmonella and cholera) as a food source. Imagine growing anaerobic bacteria – which might grow in the oxygen free atmosphere of Mars under controlled conditions – in huge vats, to be used for food. What's for dinner? But perhaps this would not deter a Mars advocate. According to Jean B. Hunter, professor of biological and environmental engineering at Cornell University – which also gave us astronomer Carl Sagan, a Mars advocate charter member to be sure – who has worked on projects directly relating to what future Martians may eat, "algae or anaerobic microorganisms wouldn't have much potential as raw materials for crew diets. They don't taste very good, they contain high levels of nucleic acids, and the cell walls are difficult to digest," via email. Meaning future Martians will have to create Earth-like living conditions on Mars to grow whatever plants they'll be using as a food source. *But perhaps this too would not deter a Mars advocate.*

PLANTS GROWING ON MARS IS A PARADOX – PERHAPS THE ULTI-mate paradox to be found within the Mars advocate ideology. Plants need both oxygen and carbon dioxide, but there's way too much carbon dioxide (toxic levels for plant growth) on Mars. There are trace amounts (about 0.1%) of oxygen – but it wouldn't even matter if that was enough for plants to grow because of the toxic levels of carbon dioxide. There's also a shortage of nitrogen in the top layer of the Martian surface, another critical requirement for plant growth – all these circumstances, including almost constant subfreezing temps, combine to form one big problem that would obviously not be resolved by simply fertilizing the "topsoil." So would scientists breed plants to grow in the existing freezing Martian atmosphere (bearing in mind the oxymoronic "anaerobic plant") – or the terraformed Martian atmosphere, which would have a ratio of gases no one could possibly predict or control, and would be changing constantly? Even in theory, one strategy is more impossible than the other.

And whatever edible raw materials that can be grown on Mars will probably *not* be processed to look or taste like their favorite Earth foods, so it's probably in vain that the biological and environmental engineering department at Cornell University, among others, would be working on "delicious and nutritious menus" for this purpose. All the extra steps to do this – converting raw but edible into delicious and nutritious – would be, like many other things on Mars, easier said than done. On Earth, we rely on factories to do this, but Mars will have no such infrastructure. Twenty-first century man enjoys a standard of cuisine that is quite unlike even that of just a hundred years ago, with a heavy reliance on synthetic and processed – this will not be possible on Mars.

> **The menu has become quite extravagant on the ISS, and Earth-like – but that's because the food is flown in on the Elon Musk "lunch cart" – which would of course be impossible to achieve on Mars, where the cuisine will most likely be lean and sparse, if not primordial – where the rule of the land will most likely be eat to live rather than live to eat. Just keeping food thawed may be the biggest concern of all, since Mars is, after all, a more or less frozen desert.**

Even lab grown hamburgers, already the subject of research for a starving Earth and using stem cells, may be too big a dream for Mars – anything involving labs may be just too high-tech for a more likely low-tech Mars.

But let's not understate the moment the first seed might germinate on Mars, as unlikely as this is. And so might begin the first page of the first chapter of the first book in the first library of a new world.

> **Does that give you a goose bump? Does it make you tremble? Does it make you absolutely shriek? When was the last time you heard anyone say "Let There Be Life?" Dare anyone say it! Neil Armstrong's famous line withers in comparison.**

Inflatable greenhouses have been proposed, using such high-tech materials as polychlorotrifluoroethylene (PCTFE) [MarsSociety.org]. If only man could eat the polychlorotrifluoroethylene. A greenhouse made from this would need air locks to maintain the interior pressure so workers could enter without the roof collapsing. Wait till the first dust storm. Imagine rockets filled with polychlorotrifluoroethylene headed for Mars. Bon voyage!

HYDROPONICS

There are, of course, other ways to grow things than in the ground, such as hydroponics, about which NASA has been doing research and specifically for its use on Mars. Hydroponics is growing plants without dirt, where the plants instead are in containers, such as wood or plastic, and supplied with water and nutrients. Hydroponics developed around the revelation, made in the mid 19th century, that plants don't need to be planted in dirt to grow, and has thus been developed to a degree of sophistication, although confined primarily to indoor pot growers and other niche markets, and where the water supply or quality is inadequate. It's worth emphasizing that no one relies solely upon hydroponics, and large-scale hydroponic farming is somewhat of an oxymoron. Yet, this may come to be – if only on a small scale, for a brief moment – in man's first experiment with growing plants on Mars – where it may not work at all, but of course the Mars advocates have their fingers crossed.

Hydroponics uses as little as 1/20 the amount of water used with regular farming, but this advantage may be moot on a planet as severely desiccated as Mars. Even on Earth there are practical limitations to the use of hydroponics – it is, after all, much easier to just throw seeds onto the ground, rather than the arduous rigging of plastic tubing to run "formula" to row upon row of plastic containers – easier, in many cases, to just ship food to where it's needed. Hydroponics would never actually be *convenient*, on Earth. So it seems this would be a more massive undertaking than one might think, for such a vision to be realized on Mars.

There are many curses that must be lifted before a Martian colony could succeed – what the Mars advocates need isn't engineering marvels but magic – which would describe the miraculous achievement of growing anything on Mars – something *edible* – in quantities that would provide sufficient calories, sufficient nutrition, and satisfy the hunger of a colony of humans – or it may be impossible. Just look at how much of our own vastly sophisticated infrastructure is dedicated to growing, distributing, and storing food – and still millions go hungry – 20,800 will die of starvation today! – that's the daily average for 2013 [http://www.statisticbrain.com/world-hunger-statistics]. "We have the ability to increase the yield of small farmers in places where food didn't grow 50 years ago," says Ertharin Cousin. She's talking about Earth! (as head of the U.N. World Food Programme in 2013).

> **Worldwide, 1 in 7 people goes hungry or is malnourished. 1 in 7. We can't feed the people on Earth! But the Mars advocates expect to feed everyone on Mars – a planet that has already given up its ghost!**

If the colony could get just one plant species to proliferate, they could perhaps grind it down into a malleable yet unobjectionable paste, which could be shaped and formed. **Get that dial-a-flavor spinning!** Even if just 1% of Mars could be made to grow something...hope the colonists aren't limited to the western edge of north facing craters – or some other bugaboo.

COOKING

Cooking on Mars would be quite different from on Earth, where water boils at 212 degrees F, because of the great difference in air pressure on the two planets. It would boil at a much cooler temperature on Mars with its much lower air pressure – as cool as 50 degrees F and lower – but only at lower altitudes where the air pressure is the greatest. All the water in a pot would boil away and the food would still be cold! – which of course has far-reaching implications regarding food preparation. You'd need a pressure cooker any time you wanted to boil water – or for the entire kitchen to be pressurized. Who would think something as simple as boiling water would be so difficult – or impossible! Other tricks would need to be devised also, as we create alternatives to the use of boiling water in the kitchen. See my remarks on the triple-point of water for a more thorough understanding of this in the respective chapter.

Vitamins

One could wonder why the supplement industry is as big as it is, as if our own abundant food supply were inadequate – certainly the need for supplementation on Mars would not be less. Imagine rockets headed for Mars, cargo ships filled with nothing but vitamins!

Dial-A-Flavor

Whether or not anything can grow or live on Mars remains a huge unknown – but we do know right now that nothing is growing or living on Mars!

THE "JOY OF FOOD" IS ANOTHER ONE OF THOSE QUALITY OF LIFE issues that we take for granted on Earth, but which would be mercilessly carved away in a life on Mars – and replaced by unending menu fatigue.

All food that is now prepared for space, i.e., for those on board the International Space Station (ISS) is, of course, prepared on Earth. This allows for a healthy – and satisfying – variety. But the ISS is only about 250 miles away from Earth as it floats in Earth orbit so it's easy to maintain its food supply – and other amenities. This same convenience – and what an enormous convenience it is – will not be afforded to a Martian colony, to which regular rocket shipments, by which I mean convenient rocket shipments, will not be possible. You can't just have monthly rocket launches to Mars carrying food and supplies, as with the ISS. Even though Mars is in our solar system – the next planet over – that's still 54 million miles away at its closest and *that is not convenient.* Given the nature of the orbits of the two planets, rockets can launch from Earth only once every 26 months or so. Even if rockets were 10 times faster, that would give us a bigger launch window, but still only occurring once every 26 months or so. Even if you could launch several rockets to Mars during these windows, the logistics at the Martian end, involving multiple craft landings in the same exact area over a matter of weeks, would never be "safe and ordinary." Every 26 months, a colony on Mars would be in crisis mode, where the failure of a single spaceship landing – whether through a crash landing or a malfunction anywhere between Earth and Mars – could easily spell the end of the colony.

EARLY MAN, LONG BEFORE WE LEARNED HOW TO CULTIVATE crops, was opportunistic. Early Martian will need to be also. As with a primitive tribe in the Amazon, who eat whatever is edible – that's within reach, from their "garden" – early Martian will be lucky if he can subsist on whatever will grow out of that red ground. Forget about Earth's lush variety of flavors and textures – we take our cornucopia of earthly delights for granted.

> **Edible won't be just the immediate goal, but the extent, where taste bud fatigue will be a given – and might just drive you crazy.**

From the beginning the Martian colonists will need the ability to produce a surplus of food, but what if the first "growing season" on Mars is a bad one? – even on Earth, some years are better than others. But the colonists would obviously need to create a surplus of food right away, and they would need a 100% reliable and predictable reserve of food for the initial colony as a prerequisite for a larger one. It seems this would require a lot of experimentation, experimentation that could only take place on Mars – this would clearly be the ultimate test because if a colony can't be self-sufficient, there can't be a colony.

The "farms" will need to be heated, if only to keep the temps above freezing – but this would obviously be a *tremendous* challenge on the subfreezing Mars! – involving either huge above-ground structures or huge underground caverns with artificial lighting.

Whether or not anything can grow or live on Mars remains a huge unknown – *but actually we do know right now that nothing is growing or living on Mars!* Furthermore there have been no simulations that replicate in their entirety all of Mars. We would need to grow not just plants, and not just plants that are edible, but plants that provide sufficient calories and nutrition. There's an art and science to growing food – discovering and culturing plants that we enjoy eating and that are nourishing – and there would need to be a multitude of plant species to achieve this. Can a horrible lack of variety that seems inevitable, and with it a complete denial of the palate, be avoided?

Mesquite grilled albacore steaks are served on the ISS but we will discover the hard way that one can't extrapolate from the ISS what life will be like on Mars.

Generally, anyone moving to Mars will be forced to give up things that even those on the ISS enjoy – including not just access to a reliable supply of clean water, free of contaminants, but every beverage known to man – many of which have been crafted through centuries of art and science – beers, wines and other alcoholic drinks, milk, soft drinks, flavored waters and sport drinks. You like chocolate? There won't be any on Mars. Add to this the infinite variety of cheeses and breads, edible oils, nuts, candies, specialty and gourmet foods from every country... none of which will be on Mars, ever, and since most food items have a high water content, and water is heavy, and the cost of shipping is based on weight, not much from this long list would be shipped via rocket. And you can't simply make a powered version of these things and then add water – some things, but not for the most part. Did somebody say "jobs program?" Oh, you think industries will spring up to supply a Martian colony with freeze-dried versions of the foods that we eat on Earth? Revive and invigorate the economy? It certainly wouldn't make sense to exert such an enormous effort to feed a tiny population on Mars – which might never exceed one or even a few dozen colonists – while at the same time *tens of millions* go starving on Earth (and not for a lack of creativity) – that irony would be insufferable.

Coffees and teas might be transportable – but our addiction to caffeine will probably have to be restrained on Mars – since it causes increased water loss (by increasing urine flow) – which would obviously be a big concern on a desert planet with no readily available supply of water – and as a drug may have other side effects particularly undesirable on Mars – and consequently may be administered in carefully controlled doses rather than as a routinely consumed beverage. By carefully controlled doses I mean people on Mars would probably not be drinking coffee or tea by the gallon like we do now. Sure, they drink coffee and tea on the ISS, but again, this and other *amenities* enjoyed on the ISS are afforded by its close proximity to Earth. I believe we will discover the hard way that one can't extrapolate from the ISS what life will be like on Mars, regarding the food supply. It's probably better to think that Earth is to the ISS what the ISS would be to Mars. Because going from Earth to the ISS is a downgrade, obviously, but in going from Earth to Mars, one can expect an even larger compromise – and for which there would be no compensations. This does not spell F-U-N.

The degree to which some – many – of us earthlings crave variety in our menu might even be considered a maladaptation. But this would be different from the *capability* to eat anything, which would of course be an advantage over fussy eaters who might starve in an "eat to live" environment – which would be the case on Mars.

One possible exception to the rule that severely limits the shipping of high water-weight items to Mars would, of course, be blood products, but which present a unique problem due to the refrigeration requirement, so it will be interesting to see what along these lines will be included in the cargo of the first rocket to Mars. NASA spaceships haven't generally had refrigerators since all the food is freeze-dried and regular refrigerators don't work in space. Blood plasma can be freeze-dried, and Funakoshi, Iijimi, et al developed a technique in 1997 to freeze-dry and manufacture blood cells, stem cells, and platelets, something generally considered to be very difficult due to the very fragile nature of red blood cells. This would obviously be an area of concern to anyone headed off to Mars.

Other Animals

Imagine "Nick's Chihuahua Alfredo Kitchen" on a sign above your door.

BRINGING PETS OR OTHER ANIMALS TO MARS WITH THE INTENTION of breeding them as a food source poses special challenges. For one thing there could never be a Noah's arc going to Mars. What would you do, put all the creatures in custom made spacesuits? On Mars all creatures would have to live in pressurized habitats because individually tailored space suits would not be possible. But either scenario, in suits or pressurized enclosures, is equally implausible, and laughable. Exceptions may exist for very small creatures like hamsters or rabbits, which don't need a lot of space to roam around, there may be possibilities here. These could conceivably share living space with humans. Imagine a significant portion of your tiny studio apartment being used as a "hamster processing center." Hamster stew. "Nick's Chihuahua Alfredo Kitchen" on a sign above your door. That might actually work in the very short term, for a colony of a hungry few dozen, but it seems it wouldn't lend itself to a thriving human society. Is this what the Mars advocates are imagining?

It's in the Details

**Would people born on Mars worship our gods,
or would their gods be...the first colonists?**

THERE ARE MANY ASPECTS OF LIFE ON EARTH THAT WE TAKE FOR granted and which have become a large part of the human condition. These are each part of the larger question of whether man can colonize Mars and none can be denied. Nor can we let the PT Barnum of our day – Mr. Elon Musk, through the guise of a Mars tourism industry – distract us from these decisions. We can't simply bring all these things with us, though that would be tempting – and could result in the creation of a microcosm of the very thing we are escaping!

CLOTHING

I expect there would be more than just "the spacesuit," so can I bring a lifetime supply of my favorite khakis with me – or expect regular shipments from Earth? What if my size changes? Jeans are rugged, durable, and warm but even this practicality may not translate to Mars, where washing clothes would be a problem, giving more lightweight synthetic fibers the advantage, so goodbye to "jeans and a T-shirt" – this could be a deal breaker for some. Realize that such supplies would be tailored to the specific needs of the Mars population because of the huge cost of shipping – which means there wouldn't be a lot of extra sizes on hand. There wouldn't be a surplus of inventory, at least not for the first few years or so – which could make things pretty inconvenient for those expecting to replicate their earthly lifestyle. My wife has more than twenty pairs of shoes. Will she have to leave these behind? I said WILL SHE HAVE TO LEAVE THESE BEHIND! The indignity. I said to her honey, they'll build a shoe factory right on Mars, there will be styles *unique to Mars.*

We will need to bring 10 sizes of spacesuits for each child that will be born on Mars – or some other number depending on how precise the fit must be – trading each size for a larger one as they grow. And age appropriate toys and teaching materials for the kids as they grow and develop and learn – and who will, of course, inherit their parents role as *guardians of the human race* – which might give new meaning to the ideas of predestination and totalitarianism.

MEDICAL AND HOSPITAL CARE

Members of the first crew on Mars will need more than just training in first aid. They will each need extensive training in surgery, anesthesia, and child birth – much of which may have to be reinvented for life on Mars. Perhaps the first colonists should be doctors and dentists, specialists in human physiology along with many other things. Such will be the minimal requirements for the first Martians, just to live at the subsistence level, where overall living conditions will be less than what passes for third-world here on Earth. You'd have to ship operating room equipment in the first few ships to Mars, with the knowledge to use it. One of the first schools on Mars would need to be a medical school, with a focus on priorities like food science and health issues that will arise but that could not have been anticipated here on Earth. I use the term "school" loosely because starting out there wouldn't be buildings or classrooms like on Earth.

MEDICINES

We take for granted the layers of infrastructure that are implicated just for this. Without well maintained stockpiles of every known medication known to man on hand we may as well be living in the Middle Ages. Are we ready for that? We can't assume that existing cures and remedies that work on Earth will "just work" on Mars, nor could Earth scientists offer any meaningful assistance with epidemics that break out on Mars. It would be a long time before Mars based facilities could support the research, development, and manufacture of medicines and drugs tailored to life on Mars. We wouldn't have the means to respond to many life-threatening circumstances, not at first (which is the most critical phase), nor any number of non life-threatening but high-maintenance maladies that could easily lead to the instant death of a small, trial-size colony. On Mars, there's no 911.

Would something as ordinary as Aspirin work – and at the same dose? Will it be more or less stable in the Martian atmosphere – ditto the remaining infinitude of drugs and medications that are a feature of modern society. Will the astronauts be subverted once the Martian reality hits them and, for example, abuse drugs? Become careless due to hypokinesis (abnormally slow movement) and die from some trivial accident or misjudgment? If I pick my nose, and it bleeds, could I die?

Ordinary things like contact lenses...the technology to make even regular eyeglasses may be a long time coming on Mars. Or just regular window glass – we certainly can't have buildings without glass windows, that is not in my vision of the future – or the artists busy with their artistic renderings. Can we expect to give up any or all these things, forever, in our new life on Mars?

TOILET PAPER

There is nothing more mundane than toilet paper, but on Mars nothing is mundane. It's relatively lightweight but bulky so would be hard to ship to Mars. Of course they use it on the ISS but that's a completely different story because the ISS gets regular concierge services from SpaceX rockets – this will not extrapolate to Mars. And of course there are no trees on Mars – wood being the main ingredient in toilet paper – nor would there likely ever be (see chapter on the basic facts about Mars, "You Like to Breathe, Don't You?"). Even if through some miracle we could get small plants to grow (underground with artificial lighting), scaling this to trees would be impossible – which means we wouldn't have any of the other wonderful things made from wood either. Maybe pressurized greenhouse domes? – which seems physically impossible –

all for the sake of toilet paper? And what will they do in the meantime, while they're waiting for the trees to grow? There's a lot of these "what will they do in the meantime" questions, between first stepping foot on Mars and an "Earth-like" colony, where things will range between miserable and unlivable.

Burning used toilet paper, along with other waste, might at first glance be an expedient form of waste disposal on Mars but, whoops, there's virtually no oxygen on Mars so nothing will burn (which also means no metals or glass industries). If we adopt a primitive lifestyle on Mars, as I've suggested elsewhere in this book, we might not need toilet paper – after all, our cavemen ancestors didn't use toilet paper – nor do all the other animals on Earth, so why do we? It seems the need – or custom – of using toilet paper is a modern *maladaptation* that may not be supported on Mars. So what will they do with all the used toilet paper? And if there's a clever way to avoid the piles of used toilet paper I'm imagining dotting the "Junkyard Mars" landscape...why are the *engineers* withholding this cleverness from the rest of us?

ECONOMIC SYSTEM

What economic system will Martians use? Will it be capitalistic – based on the U.S. dollar – some other capital based currency – or a newly invented one more appropriate to Mars?

What's the payoff for those who've given up all of life's earthly pleasures – to endure the unending and extreme hardships completely violates the pleasure/pain principle – and which will entail, among other things, living by a system that completely challenges the principle of work for pay. On Earth, of course, people get paid to work. You work for some period, get a paycheck, then pay the bills. That paradigm is almost as fundamental as gravity and the speed of light. When people do work they want the money. So, will the first guys on Mars be getting a regular paycheck? That would of course presuppose an infrastructure that includes banks and other services and businesses – an economic system – *which obviously doesn't exist on Mars!*

So it seems life for the first Martians will be based on some combination of slavery and altruism. Indeed, the Mars economy (for lack of a better word) will not be based on supply and demand as much as life and death.

Will earthlings on Mars embrace deficit spending, or a balanced budget – or neither? Would a planet with only a few dozen or hundred inhabitants even have an economy, or would it exist as a primitive tribe in the Amazon, who live simply to take care of one another and where the concept of money doesn't even exist, much less the elaborate and contrived principles that extend from it – and which will, in all likelihood, never exist on Mars.

The idea of personal property will probably carry over to Mars, for things carried with the astronauts to Mars. There wouldn't be any retail stores, which means when something breaks down or wears out, *it's gone*. When this happens, they'll have to share each others stuff, which will force them to create a system of trade, which could result in a form of currency being developed. In anticipation of this, should astronauts take their Earth money with them? At what point will Martians insist on the right to own land? And what to do with real property and assets left behind on Earth?

The gold in Fort Knox is a significant store of value for our currency. Will this carry over to Mars?

If a planet can possess the store of value for another planet's currency, could we use a light-years away diamond planet for ours?

Will there be unemployment on Mars? You think some people won't get tired of their job and quit? Might a mutiny occur before the rocket even gets to Mars?

LAW AND ORDER

Will the social fabric weave itself into something strong enough to hold everything together? Or will there be nothing but loose ends from the outset, with the only outcome – the inevitable outcome – being a colony ripped to shreds?

How will crime and punishment be addressed? When will the first rape occur on Mars? The first murder? You think people won't break the rules on Mars? Will there be a prison on Mars?

Will we bring any Earth laws or will we start from scratch? Will there be marriage? Divorce? Do we bring EEO hiring guidelines with us, or a strictly enforced merit based system – there's a lot riding on this, with little tolerance for the cute and contrived laws we have here on Earth.

Will we need police? At what point will there be a division in the ranks, leading to civil disruption or even wars? Will we end up with the same problems on Mars we now have on Earth – somewhat defeating the purpose of going there in the first place!

GOVERNMENT

The political nature of our American democracy, with its intrinsically short-sighted 4-year presidential election cycle – which antagonizes some examples of long-term planning that would benefit a Martian colony – may not therefore lend itself to life on Mars. So should Mars have a monarchy? A benevolent dictator? A single, long-term plan is obviously essential – this rules out the political system we use now and its disruptive 4-year cycle.

WOULD THE CONSTITUTION BE ON MARS?

Not if by this one means the Constitution of the United States. There would be a government, and laws – eventually – but you don't need a constitution to have a government. There wouldn't necessarily need to be a constitution on Mars at all, certainly not at the beginning. All countries do not have a constitution – a constitution does not guarantee an effective government. Items in the American Constitution in particular would have absolutely no bearing on a human colony on Mars, which would be international – but even the idea of "international" would become meaningless on Mars because of the singular nature of any colony. The American Constitution was designed to address particular issues that do not, obviously, exist on Mars and if there were ever to be a Martian constitution its intent and scope would be entirely different from the American Constitution. Any colony on Mars would have an intrinsic sovereignty. If at some point a Martian colony wanted to formally declare itself independent from people or governments on Earth, a constitution might be one way to establish this independence, but simply by virtue of the inaccessibility of the two planets, it would probably never be necessary to make this point explicit. One cannot move to Mars without giving the strong message that one wishes to have nothing to do with Earth! – and simply by stepping foot in the rocket, one is absolutely proclaiming his independence from Earth! It is hard even to imagine that either planet could ever have any kind of jurisdiction over the other or its inhabitants. There are those who question the relevance of items in the American Constitution even now, but its relevance to a colony on Mars would certainly be zero. This is equally true for all the other established philosophies, religions, belief systems, doctrines, principles of conduct and behavior, etc., that exist on Earth. Mars is a completely different world; one can not simply presume to make analogies regarding law and order a priori.

RELIGION

Would the idea of religion on Mars, starting out, even make sense? All the mythologies of all the religions are only relevant to us as earthlings, and on Mars there wouldn't be the numbers one normally associates with an organized religious group. Would the Bible have meaning on Mars? The birth of Jesus would certainly have no meaning to anyone born on Mars, and therefore also Christianity and Christian ideals – but these are the foundation – and inspiration – of much of our civilization, and what gives meaning to the daily lives of many the world over. Would people born on Mars worship our gods, or would their gods be...the first colonists?

Do we completely abandon all religious dogma and start from scratch – after all it seems there is an inherent religiosity in man that will take one form or another – but will this also be true on Mars, where man may revert to some primitive state that exists only to survive, a cretin or moron able to manage the hydroponics but otherwise unaware and devoid of culture.

If you die on Mars, is redemption possible? Can God watch over people on both Earth and Mars? These questions are, of course, more relevant to some than others.

Will we bury the dead in cemeteries, or recycle them to reclaim their water, as in the novel *Dune* – which doesn't, thankfully, go into the details, and neither will I propose the details, but the question demonstrates how real things are going to get if the Mars advocates have their way, who'll be doing things on Mars that are purely disgusting – and ethically reprehensible – to the rest of us.

DOMESTIC LIFE AND LIVING QUARTERS

Will my Martian home have a dirt floor? Furniture? A desk, chairs, cabinets, shelving?   On Earth we've become accustomed to certain standards for these things, using various woods, plastics, composites, veneers, metals – most of which are too heavy and bulky to ship to Mars – but none of these things will be on Mars – there are no Ikeas or Home Depots on Mars. These are the things man will miss most as he's settling into his domestic life on Mars (besides air, water, and food, of course). These do not pose mere inconveniences. And life on Mars would be more domesticated than we might think, due to its deadly environment, where simply staying indoors would become the status quo other than for necessary work activities. We will be heavily compromised in our abilities to ship "furnishings" to Mars, since priority would be given to food, medical supplies, spare parts, and other essential things. This would obviously compromise our modern standards. But can you imagine living without these things – what, they'll just throw things into piles on the floor, like sloppy kids who won't keep their rooms straightened? Pipes for plumbing, insulation for walls, carpeting, drywall – all these things are bulky and heavy and can't be "just shipped" to Mars and there are, of course, no equivalent industries or resources to supply us with these things, or easy substitutes. (Drywall would probably not be suitable for Mars anyway, where all living quarters would need to be pressurized.)

People landing on Mars would not be able to just devise substitutes for these things out of the freezing desert ground – which is all Mars is: freezing desert ground. That's the entire planet. Mars is just red dust and dirt! It's impossible to just convert the Martian dirt into these things – but which the Mars advocates actually do expect!

Some have claimed, in published reports, that people on Mars will be able to just throw the Martian dirt into a 3D-printer to make anything – which is of course journalistic malfeasance and the authors should be thrown into jail. Do you see why I accuse the Mars advocates of being mentally ill and misled? I'm sorry to report that such magical things *will not be happening on Mars*. All this means (among other things) that the living quarters would have the ambience of a workshop or warehouse *at best* – assuming structures that are prefabricated on Earth. Colonists would in fact be living more like nomads in the desert, hobos, with dust everywhere and non-stop complaints of "it's always cold in here," etc. Do the Mars advocates not have the intelligence to anticipate these realities? But if that's your thing then step right up. Mars: a life of adventure.

The Internet

The internet can be a big distraction to many of us already, but on Mars this double-edged sword could become even more sinister.

WILL THE INTERNET BE ON MARS? YES AND NO. IF THEY HAVE electricity on Mars, they'll have computers on Mars – at least starting out, but this may be an entirely quixotic expectation on my part in light of my previously stated predictions for a low-tech Mars, especially if the colonists are living a bricklaying gym rat/hydroponic farming existence, where the need for computers would be limited to the monitoring of life support systems, rather than be the heartbeat of an existence that defines 21st century earthlings. So let's say that starting out they have computers, and that they're interconnected. Some would call this "an internet." But by internet, others are referring to the web pages that we've become familiar with, signified by the http prefix, which means Hyper-Text Transfer Protocol. Still others would be thinking of the massive collection of web sites that have redefined life as we know it – the millions of web pages and services that include Youtube, Google, and social media, and that allow us to map, bank, and take college courses, and to download or stream music and video. But many of these are nothing but conveniences, even silly, and would have no relevance to living on Mars, at least not starting out and perhaps not ever. That's right, the first colonists won't need the internet – because they will take with them all the knowledge they need on solid state or other storage media yet to be imagined, with perhaps every book, manual, and encyclopedia ever written. They'll bring with them the entire contents of the Library of Congress – in digital form – and a snapshot of the entire internet – which might seem to contradict my claim that they won't need the internet, but not really. Any dependencies between Earth and Mars would handicap a Martian colony – the internet would only be one more dependency, and stymie self-sufficiency. In fact bringing the equivalent of the entire internet with them would act more as a security blanket than as a vital source of answers and solutions, though, since man will leave Earth still thinking like an earthling, where being online has become such an integral part of our daily lives that being offline seems unthinkable. But once the rockets launch the astronauts will quickly shed their online routines. It's *fortunate* that man would not need the internet starting out on Mars. So the first humans on Mars will definitely not be surfing the internet. What would they Google, where to eat on Mars?

Humans on Mars would represent the leading edge of human knowledge anyway – regarding living on Mars – after all their training and studying and then *being* on Mars. Any human standing on Mars would know much more than could be provided by Google. They could obviously not use this or online forums or chat rooms for quick tips on how to live on Mars. There will be no guidebook on the internet for the first colonists on Mars! That is the only reason the colonists would need the internet – they would therefore not need the internet. Technically, the first colonists could easily set up an antenna to receive the internet – but it would be pointless and a waste of time. Also, signals coming from Earth would be interrupted for about half of the day due to Mars' rotation, which would hinder any internet dependency (although this wouldn't be a problem with telecom satellites orbiting Mars).

The internet both fuels and is fueled by our information driven society, which is in turn a consequence of what has come to be our machine-based and automated existence that is as much defined by overpopulation and a culture that is leisure and recreation oriented. But a colony on Mars would have none of these and would especially not be information driven. Nor would a colony on Mars be a scaled down representation of us. Instead, a machine-less existence may be the only option for a successful colony, as I describe more fully in other chapters. This would certainly diminish even further the relevance of the internet. Even if we go in the opposite direction, towards a more technical Mars colony, where early Martians are able to achieve what is equivalent to a pre-Industrial Revolution existence, they would not benefit from the internet, it would be anachronistic. Early man on Mars would simply have no use for the internet.

With only a few people on Mars there wouldn't even be a Mars-based internet, as distinguished from an Earth-based internet – and it seems this would always be true. However, the ability to send some combination of news and information between Earth and Mars would remain important – even to a Mars that is out of sync with Earth, as I contend elsewhere. For example, mission control may transmit data to the colonists using other protocols, or documents that can be viewed using a normal web browser, but these would not be the same as "the internet."

Some may insist upon a more futuristic vision. Mars might some day tap into our internet, but any role played by our internet on Mars would be much diminished from its role on Earth, especially regarding internet based services and businesses among many other things, so this partic-ular usefulness would be nonexistent. Although real-time transactions would always be impossible (that problem with seamless communica-tion again), one could easily envision regular data dumps from Earth, which the colonists could browse through in some sense of real-time, and which would at least provide them with a supply of fresh reading material, although a constant-on connection would be unfeasible (and only one central computer would have access to our internet anyway) so streaming media, such as with internet radio, would not be likely. Media entertainment – which one can fully expect to be on Mars no matter how dismal things get, rather, because things may be so dismal – would be provided by massive archives of digital music and movies brought from

Earth – assuming, of course, more vital priorities like food and staying warm don't prevail. But of course these other priorities would prevail! It seems that basic life support issues would always be a concern on Mars – there may be no time for entertainment at all. Perhaps the whole idea of digital entertainment would be ill-conceived.

Eventually, internet based communication may be the only way for Martians to stay "in touch" with Earth, if somewhat reinvented for use on Mars. The internet (i.e., the technology used for this) would actually lend itself quite well to this purpose, especially since it has no moving parts and its basic infrastructure is relatively simple, lightweight, and easy to maintain.

The usefulness of the internet on Mars, however, would also be limited by the extent to which it creates "Earth envy." The internet can be a big distraction to many of us already, but on Mars, this double-edged sword could become even more sinister.

25

Requiem for Mars

I hear a requiem lightly in the distance, for Mars, long dead – and it's getting louder.

WE'RE HALFWAY THERE. HALFWAY THROUGH THE BOOK, THAT is. Have you seen anything that's on the plus side for Mars? Are you still in a rage to live on Mars?

In the same way scientists look at things in different kinds of light, I'm looking at the Mars question...in different kinds of light. In each chapter I look at the question in a different light, so even though the question remains the same, the answers can be completely different – and more fully expose the Mars advocate mentality.

The case against Mars seems to be unlimited in both scale and scope. I'm actually a secret Mars advocate but we're racking up a big negative score for Mars so far, with absolutely nothing to assure a skeptic that a Martian colony isn't just doable, but the next logical step in man's future. So let's keep searching for something on the positive side.

For every question I ask or point that I make there seems to be ten more, no single one being a *proof* that man shouldn't go to Mars, can't live on Mars – but that would be too easy. Some of the arguments I've put forth are strong, some may be weak, some more metaphysical or philosophical but the sum total, the gross weight – the *aggregate force* – can't be denied.

I'm offering as many arguments as possible (to deprogram the Mars advocates); perhaps only one will be convincing and resonate with you, the reader. I am, of course, appealing to a broad audience and can't expect any one argument or anecdote to convince everyone. I may even have insulted some of you by calling our species simpleminded but of course this may be unavoidable, up against the wire as it is, the spotlight set to max and unblinking.

The preoccupation with Mars may be construed as a preoccupation with the survival of our species (or more bluntly, reproduction and raising families – which is a pretty simple thing) but our simultaneous preoccupation with the *individual* does present a conflict of interest. It seems we might "get to the stars" sooner without this conflict, but our priorities are what they are. It is also true that our prospects on Earth may be daunting – but are downright attractive, in terms of species survival, compared to the nonexistent or imagined alternatives pursued by the Mars advocates. Earthly disasters may wipe out millions, causing untold weeping and inconvenience – but this would be true for any planet, now or ever, close or far, including Mars.

Our species is really not in jeopardy, given our incorrigible abilities to adapt, build, and engineer. There is no perfect planet, certainly not Mars. If species survival is the singular intent, Mars is neither safer than Earth, nor safe in any sense. Any plan to move to Mars is out of a twisted, self-serving capitalism and a fool's mission rather than any sense of what's good for mankind.

26

Terraforming

There is – in fact – no science behind terraforming or we'd already be doing it here on Earth to offset the effects of global warming. Scientists slaphappy about terraforming should apply the principles here and now, convert our desert states to verdant grasslands and reverse global warning – it seems perverse to give all their attention to another planet where there is nobody and nothing! Terraforming has no real meaning, other than as science-fiction. It's a weird idea that violates the general rule that living things adapt to their environment and not vice-versa.

MARS ADVOCATES SAY "MARS IS HOSPITABLE." WITH THE NEXT breath they say "Mars will require 1,000 years of terraforming." Well, which is it? Both can't be true because if Mars is already hospitable there would be no need to terraform it! – the two ideas are mutually exclusive of each other. How can the Mars advocates, while insisting that Mars is already hospitable – *hospitable!* – at the same time provide much argumentation in favor of terraforming? Am I the only one to notice this – glaring – paradox? As I indicated in another chapter, hospitable means San Diego-ish, which has average monthly temperatures of 70 degrees F year round and *where no one wears spacesuits*. It's true that the many articles that describe Mars as being either hospitable or habitable go on in detail about terraforming. It seems this contradiction alone would ruin their credibility – they can hardly accuse me of sabotaging their dream, they're doing a good enough job of it theirself.

This is just another of the most obvious fallacies, or contradictions, surrounding the imagined colonization of Mars. As if 1,000 years of terraforming was something trivial. Terraforming has been proposed as a way to make Mars more like Earth, and therefore more livable, most significantly by adding oxygen to the atmosphere. But the efforts to do this would be anything but trivial! – and which underscores the contradiction and therefore also the Mars advocates ridiculousness. To emphasize how nontrivial terraforming would be, as it pertains to the Martian atmosphere, this would not simply involve "tweaking" the atmosphere by, for example, a few degrees, but to *completely change the chemical composition of the atmosphere of an entire planet!*

In fact on the ridiculous pyramid, terraforming is the apex. No, it's actually much worse than that. This is actually difficult to explain, because there is nothing even approaching terraforming that has ever been done – it's a set of ideas loosely based on conversations and the idle chatter of engineers (those *engineers* again), which means there really is no such thing. In an interview, astronomer Dr. Pamela Gay said "you could conceivably go out and start capturing comets. Comets are rich in things like water that you can tear apart and turn into atmospheres." This is presupposed on a theory to explain a planets gaseous elements and hence its atmosphere. But the time frames involved in this, as it might occur naturally, are millions of years! Another (slaphappy) theory of the terraforming mindset proposes to *melt Mars* – with a giant, orbiting magnifying glass! – then "wait around" while it cools. Yes, this is what scientists talk about! Those mad, idiot savant scientists. And someone's *paying them.*

If terraforming existed, we could use it right here on Earth, for such things as making it rain in Texas or stopping the hurricanes that continuously threaten the states bordering the Gulf of Mexico – which would seem easy compared to making it rain on Mars! And if we can't pull off such tricks on Earth, where we have the home court advantage, we certainly couldn't on Mars, where we'd be unwelcome visitors.

The reason terraforming doesn't exist in the first place is because Earth is already habitable – whereas theories of terraforming have been invented for the express purpose of making a place that is uninhabitable, habitable, and applicable to Mars – although there are some very recent attempts at what might be considered terraforming (or *geoengineering* as it pertains to Earth), but which usually occur "under the radar" because they are so controversial – or illegal, like the case of the rogue American businessman – *business* man – Russ George dumping 100 tons of iron dust into the Pacific Ocean, off British Columbia, during July 2012, to promote algal blooms, which would eventually sink, taking their carbon load with them (10/15/12, http://www.guardian.co.uk/ environment/2012/oct/15/pacific-iron-fertilisation-geoengineering). Carbon load? But of course. But even on a large scale, tweaking the carbon content of Earth's atmosphere seems too trivial to serve as a good example of terraforming, especially since it mimics existing natural processes. And completely useless if "global warming" turns out to be part of a natural cycle, as many argue.

> **Schemes to terraform Mars wouldn't simply speed up existing processes but disrupt them entirely, or be in contention with naturally occurring circumstances and eventually be undone.**

Which is why they're so controversial. True terraforming has never been done and remains science-fiction.

Could it be that Russ George is actually a scientist – but claims to be a businessman to avoid the label "mad scientist?"

Some will argue that our current technologies are serving to geoengineer the planet, if in some accidental way – which is a fallacy because if it's an accident then it's not engineering and therefore not terraforming. The point of engineering is to deliberately create some specific effect, based on scientific principles – not hope for some inadvertent but "beneficial" effect, which would be an entirely different thing – serendipity – which is not the tool of scientists or engineers. The use of artistic license to call such a thing geoengineering or terraforming only confuses the line between science-fiction and science fact.

Terraforming Mars would not be a matter of so-called megaengineering projects as in creating levees, stadiums, bridges, or skyscrapers

– that would be another misconception. These would be puny compared to the goal of terraforming, which refers to literally changing the nature of things on the scale of a planet, in particular the atmosphere – to change the chemistry of these things. Grooming a beach is not terraforming – it was a beach before and after, it's simply prettier. Some may have even preferred it in its natural state. One should avoid trivializing the idea of terraforming Mars by thinking it is simply a large-scale landscaping project. Terraforming Mars means to literally change Mars into another Earth. *To create another Earth!*

Terraforming is not just a vastly scaled up version of green suburban landscapes in the desert complete with 18-hole golf courses. Anyone can build a golf course in the desert. It's simple enough to either build a ditch or lay pipes to a water source. Both before and after the grass is growing, it's still desert. That difference is purely cosmetic. Terraforming is not simply a scaled up version of this. Terraforming does not refer to small scale – or large scale – cosmetic changes to the landscape. A golf course in the desert is hardly analogous to building a colony on Mars. This is just more of that dumb engineer thinking and trivialization.

This is not to say that terraforming will always be impossible, but even here, some experts agree that terraforming would probably not be feasible for another thousand years, which makes it impossible to discuss in a practical context, and presents sufficient grounds to table the discussion of colonizing Mars for another 1,000 years. But by putting something 1,000 years into the future, aren't they essentially saying it will never happen? Evidently not. They're not being at all sarcastic when they say it will take a thousand years. But this is obviously a wild guess. To be sure, everything in this arena has *gone wild.*

It may make more sense to think that we will have discovered another water planet like Earth, and invented the means to get there, before we could put actual terraforming techniques into practice, i.e., change Mars into another Earth. Both scenarios are highly unlikely – and outrageous – but the former conception does seem a little less futile than the latter. Bearing in mind each newly discovered exoplanet is even more light-years further out, yet, ion-powered rocket ships carrying 1,000 passengers to some as yet undiscovered watery exoplanet in a distant solar system makes more sense than a plan to convert Mars.

Terraforming is like time travel – neither one exists, yet people talk about the two as if they did. We are probably better off without them. I'm glad neither of these two things exist, and I hope they never do, because they both presume to upend the laws of nature, which would be a very bad idea. No matter how one looks at it, you couldn't possibly bank the future of a Martian colony on terraforming.

27

The Arrogance of Engineers

In defying the physics that keeps us here, it may be impossible for us not to destroy ourselves – what grand creatures we must be to think we should. It may be a greater freedom to simply embrace our mortality, and the relative tranquility of what we already know.

IT'S AMAZING THAT ANY SCIENTIST COULD THINK, TODAY, GIVEN what we know, that we could live on Mars, or cultivate plant crops...to actually colonize – colonize! – Mars. Even before the rovers – before satellites – before anything that is modern – only the most imaginative would be taken by such a fantasy! But now? The lessons learned from the rovers are already old! It's as if we've gone retrograde, and decided to just IGNORE the science! All the dumb, crazy scientists creating dumb, crazy science, thinking they're saving humanity while America and the rest of the world goes to hell in a handbasket...which is happening because of the dumb crazy science cycle, or the **smart science-crazy science cycle**. The scientists are creating the very problems we need to escape from – are they all just nincompoops like the rest of us?

To think that one can make Mars more like Earth, through the ridiculous and imaginative process of terraforming, is arrogance at its most breathtaking. No! – at its most heart stopping. No! – at its most earthshaking! The arrogance of engineers, of the profession itself. With their incessant, mindless "we can do it we can do it we can do it," which may be more applicable to terraforming than anything previously. Engineers are blue collar working stiffs, not wizards. The degree to which they make our lives better is highly debatable, there is a point of diminishing returns, and that point may be long past. Is each year that passes better than the previous? Of course not. They were instrumental in the economic crash of 2008, inventors of so-called "derivatives." A few are responsible for truly useful inventions or the rare awesome gizmo, but most are dime-a-dozen technicians who can be switched out for the next guy who walks down the street after a few months of training and have zero capacity for innovation.

The funny thing about being an engineer is, a project may be a complete failure, but they seldom get fired because if an engineer can't do it, it can't be done! It's not as if you could replace them with someone from a "higher tier." At least that's the thinking. Engineers are given a certain carte blanche, whereas anyone else further down the chain would be fired. I wouldn't trust someone just because their job title includes the word "engineer," or because they have a degree in engineering, who are becoming more and more, literally, a dime a dozen. The profession needs to be reorganized so that the minimal requirements for someone to call theirself an engineer are a masters degree plus a certain number of years of experience. That way whatever expectations we may have are realistic. There are too many people, including those with engineering degrees, claiming to be engineers who are not worthy of the title and its connotations. It seems we are graduating too many of them, contrary to reports that we need more, who are left hanging by a thread and who latch onto the most ridiculous projects and ideas – like colonizing Mars and all that that entails. A learned society doesn't equate to more people with engineering degrees when the current excess doesn't seem to make this world a better place.

28

Idiot Savantist

People in the space industry are drinking the Kool-Aid full strength. They're literally paid to believe, their mortgage payments depend upon it, so their "passion" is understandable. But it's a brainwashed corporate culture nonetheless, impossible to admit that there's nothing – absolutely nothing – about outer space on the plus side for human habitation.

God forbid this mentality leaks into such a serious thing as the diamond-hard science of outer space exploration – because I don't want to be on Mars looking at my time-travelling-clown-faced baby when I could be back on Earth merrily tip-toeing through the nuclear wasteland the Mars advocates had promised.

THE "VITALITY" OF GOING TO MARS CERTAINLY ISN'T ESTABLISHED or in some sense a truism. We're approaching the point where travelling to Mars might be at least technologically possible but it seems that whether or not man ever colonizes Mars would not be a matter of rocket science any more than say, moving to Colorado is a matter of automobile science, where my *prerogative* to move to Colorado would have nothing to do with the state of automobile science or advice from car manufacturers. This might just boil down to how the rich are different. Imagine you (Elon Musk) have a rocket but don't know what to do with it. I have a rocket but don't know what to do with it! Let me see, I'll...colonize Mars. Colonize Mars? *Colonize Mars!!!* Going to Mars just because you have a rocket is like going on a transcontinental road trip from New York to the tip of Chile, South America just because you have a car and *nobody does that even though there might be a good reason to!* I'm sure you could find someone to rationalize that such a trip to Chile might even be vital!

If man, in the all inclusive sense, decides we should colonize another planet, naturally we can give these intentions to rocket scientists and engineers, but this should not be the other way around.

For example there might be a consortium, comprised of individuals from every walk of life and profession, to "weigh in" on this Mars mission venture. This is simply too big for it to lie in the hands of a self-appointed group of hubris-ridden rocket scientists or worse – the solitary voice of a veritable eccentric millionaire "rocket man" crying "all aboard!" who is obviously more megalomaniac than humanitarian.

So the rocket scientists should not be part of the discussion to begin with. All that is necessary is that we know that rockets exists, such that going to Mars might be possible.

But it seems that any discussions that might follow from this would be brief since, given what we've known for years, Mars is unlivable! – so it wouldn't matter if there were already a hundred rockets lined up at the launch pad!

Because our existence depends upon a specific and finely balanced supply of nutrients in a complex food chain, and also upon our position in an even larger ecosystem. There would be no counterpart to any of this on Mars. But the naturally occurring imbalances, deficiencies, and deformities found in our species incriminate a shortcoming in Earth's so-called abundant resources – or our inability to fully exploit them – such that evolution is an ongoing compromise...and we're handicapped already! Such an argument can be made, but the situation could only be worse on Mars, indeed considered by some to be a dried out husk of its former self, in which case the known dearth of vital elements would not suggest an Earth-like compromise but *a losing battle*. This devastates the notion that a move to Mars is, as some insist, vital.

**The human brain is wired to play, love, and create...
to seek sensation, stimulation...all of these would be
stifled on Mars. The whole point of being human
would be defeated on Mars.**

Right now it seems the only people blowing the "venture forth" trumpet are people who work in the space industries, so the merits of "interplanetary exploration" have taken the form of an agenda that relies on brainwashing the public, as with Robert Zubrin's *The Case for Mars* (which, as I state in my full review of that book in a separate chapter, may indeed be nothing but a compilation of stray notes and idle thoughts from the author's days at NASA).

OUR IDIOT CULTURE

Typically, most people – most *successful* people – know three things, how to get to work, how to do their job, and how to get home. And in most other regards, they're stupid. One might say it's a truism that to be specialized in one thing is to be ignorant of other things – there's an obvious mutual exclusivity. This means that specialists usually don't know much, if anything, outside their specialty. The problem with a lot of scientists is they are so specialized, cooped up in their labs, their cliques, their associations, and carried away if not completely warped by their sense of "mission"...that they don't see the big picture – a vital prerequisite in contemplating Mars and its suitability to be home to a self-sustaining human ecosystem. Not just any old primate/animal/ multicellular organism from the animal kingdom. Human Beings. In some cases it stands to reason, where specialization is very focused, that a certain "idiot savantism" would occur, which compounds the problem, where a sense of the big picture isn't just ignored, but denied. Engineers, or anyone, can suffer from this one-track-mindedness, where they may become idiots at everything but their specialty. They're successful, in their chosen profession, yet idiots.

**It may not at all be a matter of "pure science" whether
or not man should or could live on Mars.**

But when combined with the hubris that seems to afflict many specialists, the results can be remarkably unfortunate, which might have been the case with Steve Jobs, who of course gave us the iPhone, and who died of cancer in 2011. He was fully aware of and knowledgeable of the cancer that was in his body – but which was completely outside his specialty. He ignored the advice of people who knew better, regarding his decision to delay chemotherapy. He had full confidence – some would say arrogance – that alternative treatments would cure him. As if a remission or cure would "just happen" – it had to! – but of course it did not. Not that the chemotherapy would have inevitably saved him, still, his particular genius did not either, rather, it seems to have antagonized his condition. Even geniuses – especially geniuses – do not know everything – in fact they may be the *least* qualified to know about humans living on Mars – and that seems to be the case with the more elite Mars advocates (where specialist or fanatic may substitute for genius), most of whom have the same "it will just happen" mentality and are guided more by inspiring but trite clichés like "reach for the stars" and "Ambition!"...than by science. The same is true for the "science writers" of the numerous print publications and online who share the same weird happy-go-lucky sentimentality for Mars. Then there are the modern day mad scientists, and mad venture capitalists like Elon Musk of SpaceX fame.

The increasing specialization of our industrialized society may be one of its downsides, where more and more of us lose sight of the big picture – become idiots. Ironically, in writing this book, it's my *lack* of any particular specialization that I am exploiting, and which assures the reader of my view of the big picture lacking in the Mars advocates, the corporate Mars advocates and in particular some of the more prominent rocket scientists, who are so specialized and who clearly lack an insight that could only be provided by thinking about more of the things they don't.

We need to be wary of placing the fate of humanity in the hands of engineers, some of whom seem to be working free of ethical boundaries and with no foresight regarding our future in space. I'm suspicious of the trust we place in engineers, some of whom seem nothing but red necks with engineering degrees. Don't think there aren't a few dumb ones in the bunch, slaphappy and fearless, who would convince you that we "need" to get to Mars, whether this is to avoid some future apoc-alypse or some other misguided altruism. Of course with the help of computers, being an engineer is even easier now than years ago. The point is, one needn't be truly smart, innovative, or bright either to get an engineering degree or, evidently, become employed as an engineer. So if one of them tells you that we "need to go to Mars," don't listen. Just because someone has an engineering degree doesn't mean they have the wisdom to direct the course of humanity.

So, Mr. Musk is an idiot savant. Idiot savantist might be a better way of phrasing it because it sounds more like "scientist." But this should not be taken as an insult – it's not like I'm calling him an idiot.[1] Because there are – and were – many other idiot savantists out there, like Einstein and Michelangelo (to whom I've already compared Mr. Musk). This is another kind of idiot savantism than those made briefly famous for their piano playing skills. Anyone with one or more PhD's (or the equivalent specialization) is on the road to becoming an idiot savantist – but how could this be otherwise? Their specialization – if only as a manifestation of their neurological makeup – severely limits the possibility of a multidisciplinary outlook – hence "idiot savantist." It might be impossible to avoid sounding facetious, calling Mr. Musk an idiot savantist. But one thing should be clear: As good as Mr. Musk may be at making rockets – or electric cars – and profiting from this talent – he is equally unqualified to have an opinion on the outrageous matter of colonizing other planets – much less be the dominant voice on the matter. Calling Mr. Musk an "idiot savantist" is no more outrageous than his belief that Mars should be colonized.

Why would a man who owns a rocket factory – like Mr. Musk – propose that, of all things, man colonize Mars? The idea would be more respectable coming from anyone not affiliated with SpaceX or the space industry – or a kook yelling at the street corner. The idea should stand

---

1   Senator Al Franken wrote a book in 2004 called "Rush Limbaugh is a Big, Fat Idiot." So certainly it must be OK for me to say Elon Musk is an idiot savant. Jay Thomas and Chelsea Handler have called Elon Musk "idiot" and "asshole" on their respective radio and TV shows (August 16, 2013, August 13, 2013) in the same context – "idiot savant" is gracious in comparison.

on its own, free of propaganda, profit driven incentive or any other agenda. So when the owner of SpaceX himself proposes such a project, one must immediately be suspicious of his motives. Is it really for the benefit of *mankind* – as he claims – or personal profit? It's one thing to make rockets to launch satellites or for defense purposes, but Mr. Musk would be the absolute last person I'd trust with proposals to colonize another planet. Does he realize there's no oxygen on Mars? No water? I know this, doesn't he?

ENGINEERS ARE LIKE HUMANS

Whereas about 1 in 100 humans is actually smart, this same ratio exists for engineers, so they are, more or less, human – like you and me. The other 99 percent of humans are relatively stupid – and this is also true for engineers! Are you in shock? Which means we should stop acting as if they have all the answers – or any of the answers for that matter. They're guessing and stumbling like the rest of us. And sometimes the rest of us have better ideas.

The engineering professions actually do not attract the best and the brightest, which instead gravitate to law, medicine, and chemistry.

Some people are "book smart," but in terms of how they see the world, or envision the future, they're stupid – when they say stupid things, as found throughout the internet and science articles, like the idea that getting to Mars is vital, or that Mars is hospitable, or that some day we will trade with other planets, or a space elevator, etc. The belief that scientists are smart seems inarguable, but the ability to envision the future – or the sense of continuity between the past, present, and future – evidently relies upon a sixth sense that some people have, some don't, and it doesn't seem to be related to academic achievement or success in general. None of the things in this Stupid List is reasonable, achievable, nor even desirable – they're therefore stupid. One could argue, it seems, based on the expanding Stupid List, that we've already realized the limits of our potential, regarding human intelligence – and that we might even be sliding backwards – the evidence is all around us.

"Stupid" is an excellent word, but is widely overused and misused, so when I say that the "mission to Mars" is stupid some explanation would be appreciated. If you're employed in the space industry of course you're not going to think it's stupid! But even if it has the short term benefit of providing some jobs, that doesn't preclude me from saying it's stupid in the sense that I intend, regarding the bigger picture that most people aren't aware of due to their specialization. A lot of industries are stupid. Sure they employ people, but they're still stupid – such as when they cause more harm than good. Regarding the "mission to Mars" it's stupid in many ways, which is not at all obvious to everyone but should become more clear as you read this.

It's obviously difficult for me to think I can get away with saying something as blunt as "engineers are stupid," but with the Mars advocates there exists the paradox of smart people being stupid. Their acquired idiot savantism becomes undeniable when confronted with the stark reality of the data we've collected about Mars even before the NASA rovers. And it becomes only too clear that the guiding principle isn't saving mankind or other high-mindedness. The true mission is laid bare – jobs – grab the paycheck – and nothing else – at the expense of course of more beneficial engineering projects and intellectual capital. By skillfully attaching abstractions like altruism and nationalism to the Mars mission this purely capitalistic incentive is obscured – and to a fraudulent degree. It's galling. So how long can this fraud, this Orwellian pre-nightmare, be allowed to continue?

THERE'S SOMETHING STRANGE ABOUT PEOPLE WHO CONTINUE TO be excited by Mars. How can a lifeless planet generate so much interest? We've all seen it by now. The mystery is gone. This is all very old. On the microscopic level there may be some variety between the various soil samples, and intriguing to geologists. The terrain is a little bit different in one spot from another, slightly. And the questions become downright metaphysical: there's a rock, why is it there? But it all leads to the same conclusion: there's no life on Mars and we can't live there. It's like poking a dead person. If you keep poking will it eventually come alive? The fact remains, Mars is dead, and the presence of ice crystals doesn't mean otherwise, or two-billion year old drainage ditches. Absent the presence of creatures walking around, we look at the micro level, still nothing. So we delve into deeper micro levels, more nothing. NASA is never going to admit that Mars is a dead planet because then the funding will stop and the layoffs begin. But at some point the whole thing becomes a gigantic fraud.

> **Within this arena there's more money to be made in not knowing than in knowing. Ambiguity keeps the money flowing to academia and industry in an effort to more precisely measure...the same old facts.**

There is, of course, more to it than that, and for which one can blame our atavistic ways of thinking, our romance with the spheres that dates back to our cavemen ancestors – but which has no basis in practicality whatsoever – is nothing but a mist that quickly evaporates in the modern light. The Mars advocates, including NASA, don't want these atavistic dreams and myths that have lasted for eons – and which some may view as being part of our nature – to be *suddenly gone*. We've been asking for centuries "is there life on Mars" so it's not like we're going to just stop asking all together now.

> **But that only makes the case for engineers being idiot savantists even stronger, because it seems there would be more inspiration in knowing – than in asking the same questions over and over again, which is not only unscientific, but stupid.**

ROBOTICS

Another example of the short-sightedness of engineers can be found in the field of robotics. Robots obviously have the potential to make our lives easier and safer, but surely not past the point where they replace human workers in large numbers, a "point of diminishing returns" in this context, not progressive but regressive when it leads to reduced employment and mankind as a whole suffers rather than benefits. Which seems to be the case. Nowhere is this more evident than in the new field of evolutionary robotics, where you have robots that design other robots – so not only are workers being replaced, leading to higher unemployment, but even the robot designers will become unemployed! How smart is it for someone to design a robot that can take their place? Journalists tend to avoid the dystopic implications of a roboticized society, sticking to their brainwashed and propagandistic "isn't it wonderful" rubric, but this is one rubicon that we should avoid at any price.

The countdown may have already begun when engineers become the least trusted – and most hated – of all the professions. It seems they are nothing but geeky nudniks fooling around with their erector sets, their regard for humanity more sociopathic than beneficial. One is reminded of the proverbial "the only difference between men and boys is the price of their toys." Stop the robotization of the world. It needs to be stopped now. Are engineers all nothing but A-holes? The massive robotization of industry will lead to higher and higher unemployment. And who will be consuming all the efficiently made products if no one has the income to buy them?

> **The business world needs to rethink the very idea of efficiency and automation – as it applies to their nifty assembly lines – if the end result is a world that's gone to hell.**

Or the government needs to step in and make it illegal to create any process or system that results in the destruction of jobs. We control the future. Nothing is inevitable, and the future *can* be as magnificent as we want it to be. No one wants a future filled with robots if that brings with it high unemployment and its bleak implications. We are not the "intelligent species" if we knowingly send ourselves to that future.

> **The future belongs to people, not machines, and if engineers can do nothing but lead us to a future dominated by robots, machines, high unemployment and a demoralized society then they're even dumber than I thought!**

Rethink the idea of intelligence, rethink the idea of success, rethink everything if that's what it takes to get off this path to certain doom. We are at the tipping point. Right now. Robots are no longer cool, they are no longer cute. It's not as if there's a shortage of manpower. If the world ever goes to the robots, we are fools for allowing it.

Another example: using the regular car horn to honk messages to the owner – which might be considered a world-wide problem. Thanks for all the noise pollution, all you "smart engineers." Do I really want my boss to know when I'm late for work? Or neighbors to know when I'm getting back real late after a night of, well, never mind. What's strange is that some people actually think these "sound effects" are cute, or even some kind of status symbol. I never knew there was such a pent up mania for cars to honk at 3 in the morning for no good reason! Would it be dumb to have our key fobs emit a sound that only the driver can hear? My heart can be warmed as it beeps "I love you" over and over again. But here's the real problem: for those who've trained theirselves to ignore these noisy beeps, honks, and chirps, walking around has become more dangerous, because if a car honks at me to "get out of the way," it's not as alarming as it should be. Plus, my brain has to make a decision, and that takes time. It takes my brain a few fractions of a second to decide if I should ignore it – which is usually the case – or to *very quickly* jump out of harms way – which are two, very different things – and during that delay I could be hit by the car, and killed. Which seems obvious but not, of course, to our "smart engineers" as they insidiously train our brains to ignore the sound of car horns – *on a global scale.* (Making the problems we're, at the same time, trying to escape.) Would Mars be safer? Promise me there won't be cars honking for no good reason on Mars and I'll go.

OR TAKE FOR EXAMPLE TWO RECENTLY PUBLISHED BOOKS, *THE Case For Mars* and *Proof of Heaven.* The former written by an ex-NASA engineer (Robert Zubrin), the latter by a neurosurgeon who taught at Harvard (Eben Alexander, M.D.). Can you imagine more respectable credentials? I don't deny the success of either – each within their own sphere. (For some reason this reminds me of something Microsoft's Bill Gates once said: "Success is a lousy teacher, it seduces smart people into thinking they can't lose.") But after having examined both books, I can't decide who is crazier. Do publishers look beyond the credentials of the author – do they actually read the pages of their books?

Both of these books (*The Case For Mars, Proof of Heaven*) were grabbed up by otherwise respectable publishers, with the latter becoming a best seller. And they're both absolutely ridiculous, promoting the same old atavistic myths and propaganda – little green men/man on Mars and the existence of heaven – as if we haven't taken a single step out of the Dark Ages. On the hierarchy of literary malfeasance, these two books would be right at the top. Malfeasance means to give rise to, or somehow contribute to, the injury of others – even if this might not have been the original intent of the respective authors – who knows.

PROOF OF HEAVEN?

A full-scale analysis of *Proof of Heaven* will reveal all its implications – and which fully demonstrates the phenomenon of the idiot-savantist scientist. In *Proof of Heaven* the author claims to have visited heaven during a coma – even though he didn't die – which isn't just contrary to all known orthodoxies – and therefore heretical – but is as offensive to more modern views of heaven as something that exists purely in the abstract or multi-dimensional. (Which, I will add, is not to say that heaven does or doesn't exist.) So in retrospect, even the publishers must be embarrassed, but no, because it was a best seller. Because the nonsense of a neurosurgeon/Harvard professor is a better quality nonsense than from someone else? Must be, because I can't think of another reason. But what does this mean – that a Harvard scientist may be sufficient, but one who is in a coma even more qualified to get his point across? Or that one needs to be in a coma to be enlightened? Then we should all enter our drug-induced comas. I was in a coma and saw black – and it was super-real and vivid. There, does that prove heaven *doesn't* exist?

Here's why the doctor's story is a fraud: he could have easily claimed that during his coma he experienced "heavenly delusions." That's provocative enough, even to make a "book deal" (in a publishing industry where there are no longer any standards of authorship). Because it seems that, even as a delusion, going to heaven would be quite a trip! His delusions were certainly more majestic than the usual, assuming they were as he described in his book, without the usual literary embellishments. Although it still didn't exactly capture the sublime essence that I'd always imagined of heaven – which would be indescribable – and substantially more than a spruced up version of...where I already am. But if I had delusions, his are certainly the kind I would want, as opposed to the kind that involve alien abduction (also sought after by publishers) or possession by the devil (those too). Such "tales of the supernatural" comprise a large segment of the publishing industry, where the profit incentive is in direct competition with any vision of a learned society. But it's not as if Dr. Alexander can now cure people by touching them, or see the future, or become the world's fastest speed reader, or possess some other supernatural ability that I would fully expect from someone who had actually been to heaven – which is what he's claiming to have done literally – and who might then deserve to be lionized or... gazed upon.

No, he didn't simply claim to have delusions. But as a neuro-specialist, of all people, he should know that his experience was a delusion. He of all people knows that the brain isn't fully understood and that it "plays tricks" on us, that even *non-delusional* states are highly subjective matters of memories, perceptions, and perhaps even quantum cognition. The fields of psychology and neuroscience obviously need to be expanded upon to include Dr. Alexander's experience, let's call it "the heaven delusion," because it should be obvious to anyone who studies the brain – and suspected by anyone who doesn't – that he wasn't actually in heaven as he claims. Even if he doesn't explicitly state that he was in heaven (in his book or subsequent interviews) this wouldn't matter, since "being in" heaven or "experiencing" heaven – or whatever grammar you prefer – are all metaphysically the same, and tautological, so it wouldn't matter what words he uses to describe it. One can't experience heaven in some "lesser" way. It would be experienced fully, or not at all. This is a basic – and crucial – tenet of theology. There's no in between – you wouldn't say you saw a little bit of heaven, that's like saying you're a little bit human, so let's not mince words. If Alexander's claim is that he experienced heaven in some inferior way, then the point of his story is even more obscure – and more ridiculous.

There are two possible reactions to *Proof of Heaven*, and both of them are extreme. One is that the author be burned at the stake as a heretic. The other is that he be installed at the Vatican and be venerated as a living relic – and more almighty than the Pope. But neither of these two things happened – at least not yet. Alternatively, an argument can also be made that he was cast out from heaven, in which case he would obviously be a fallen angel – or antichrist. But in our fast paced society nobody has taken the time to think about all this.

If another Harvard neurosurgeon claimed he had a vision, while scrambling eggs, and from this vision claimed a proof that heaven did NOT exist, would a publisher give him a small fortune for the book rights? Something makes me think not. And why would some agent sitting in a dusty office be the one to decide that being in a coma adds validity to a vision? But why do we argue the matter, let's just settle back in a drug-induced coma and see for ourselves! Not that Alexander's coma was drug-induced – would it matter?

Why do "ethereal visions" during a coma – near death! – equate to a book deal and getting rich but not if I have the same visions while I'm driving my car or scrambling eggs? Why does one context charm the snake but not the other?

The religion fed to children is like training wheels on a bicycle, but at some point you gotta rip the training wheels off and let 'er rip.

Modern society continues to perpetuate an idea of heaven that's childish (one strongly rooted in Materialism) – the Alexander book is no exception. A strong case should be made that children should not be exposed to religion at all because much of our adult perspective of such lofty abstractions as god, heaven, and hell is based upon what we're told as children, and these ideas – evidently and unfortunately – often don't grow as we do. The religion fed to children is like training wheels on a bicycle, but at some point you gotta rip the training wheels off and let 'er rip. I'm not saying heaven exists, far from it, I'm just saying at least have an adult perspective. Modern man is very medieval in this regard but that's another book. In the movie *Heaven Can Wait*, which obviously has a parallel to *Proof of Heaven* (both of which demonstrate a materialistic view of heaven, as opposed to one that is purely metaphysical and which would seem more likely), a man dies at a relatively early age (mid 30's or so), goes to heaven, but convinces God to let him return to Earth to "finish his life." Now, even if the process of death were reversible, and you're IN HEAVEN – why would you want to return to Earth? The idea that you could go to heaven and then return to your regular life is too ridiculous, but let's just pursue this for the sake of argument. So, you're in heaven, you've got wings, you can fly (metaphorically if not literally). It's a 24/7 orgasm (or would be on some metaphysical level if you buy into the program – I don't but Alexander does) so *how absolutely depressing it must be to return to Earth and you're a working stiff again!* What a horrible devastation it must be, for God to tell you that, He's ripping off your wings. Perhaps we should...mourn for Dr. Alexander. It's back to that peer-reviewed nonsense and with Yellowstone about to blow! What kind of a rip-off is that? What a *cosmic downgrade!* Yet none of this is conveyed in Alexander's *Proof of Heaven,* who seems to be...quite satisfied with himself! This alone proves he's a fraud. Like someone swimming to America from Cuba and then deciding, No! – Wait, I want to go back! His book deal may be a consolation for being a mere mortal again – so we don't have to mourn for him so much. I'm not saying God does or doesn't exist, I'm just making the larger case that if he/it/whatever did, *do you think he would second-guess his decision that your time is up*? Do you think that souls in heaven are negotiating with God? Do you think if you had an argument with God that you'd win? Why would anyone *choose* such a cosmic downgrade? And back on Earth you'd run the risk again of any number of major violations which would surely send you straight to Hell! It'd be like graduating summa cum laude and saying "Wait, let's go back to first grade again" knowing that next time around you might end up flunking out. Or winning the mega-million lottery then giving all the money away. Or a giant sequoia reverting back into the seed from which it sprang. *Eben Alexander is suggesting that when I finally pass through the pearly gates that that decision...isn't final!* The idea that the gates of heaven might swing both ways, as they must for one to visit heaven but then return to Earth, strikes me as being terribly bad news. Earthshakingly bad news, because it completely contradicts the promise – of all the major religions – that heaven is eternal, that when I die, for example, I will spend the rest of eternity there. This opens up some rather weird possibilities, such as going to heaven, then back to Earth, then possibly to hell – which we can now assume is as equally nonpermanent as heaven – then maybe back to Earth again, then heaven, and...where would it end?

> **Certainly if there is a God this could not be – or that I would have the prerogative to rest in peace in some literal sense, instead of ricocheting through infinity like a ping-pong ball.**

So these implications of Dr. Alexander's experience, if one believes him, are staggering – and he addresses none of them! I'm sorry, but the metaphysical state of heavenliness would be – needs to be – incontrovertible – which is in fact a major expectation of all religions! – and far surpasses any earthly experiences, even sex – which makes me wonder why people expect to have a lot of free sex in heaven. Thinking one could have sex in heaven is the most childish thing a person can think, and it's amazing how many adults think this. It's evidently such a basic part of our nature that some think it would continue even after we're dead, but sex is what we do on Earth – *it is an inherently earthly contrivance* (more than we might think) and the metaphysical planes of heaven would be far removed from earthly contrivances of this or any other kind.

And obviously our understanding of life and death isn't complete if we think there can be an in between state – *near death* – during which claims of otherworldly or metaphysical experiences would be untouchable! Even if we don't know what the actual tipping point is, it seems a person can't be both dead AND alive, which means we're either dead OR alive. Therefore, Alexander was simply high, or hallucinating, on the drugs being administered to him in the hospital – or regardless of the drugs but through some other *weird and wonderful machinery* – which caused him to have his heaven delusions.

> **What's disturbing is that Dr. Alexander, in consideration of his specialty, wouldn't automatically know this – and make this argument himself.**

Or, of course, it could be nothing but a big lie. The whole thing would be so much more credible if only he had been scrambling eggs at the time!

HERE'S ANOTHER EXAMPLE. IMAGINE TELLING YOUR BOSS THAT you won't be able to finish some project until the day you retire (another 30 years). You'd be crazy to do that...unless you have the hubris of the scientists and engineers in the space industry in particular and convince them it's for the sake of humanity and that you're therefore unexpendable.

> **When one of them says "another 30 years" (the span of a typical career) this means "let the next generation of engineers complete what to my entire generation are impossible challenges."**

This does two things. One, it shows you how even those directly involved can't think past the current or next generation of technology – because it does, after all, involve obstacles that truly belong to sci-fi and fantasy. But rather than it being a shameful admission of defeat, it is entirely self-serving, and embraced, because – two – it gets today's scientists off the hook. But since the scientists at the same time insist their work is a vital part of the journey, which might span generations – this guarantees an extension of the funding for generations! Instant job security – without the need for any real or tangible results. In other words, a scam. Let's keep dreaming – and get paid for it!

The dreamer mentality is, of course, rampant, so much a part of modern American life, even outside the Mars advocate confines. Dream job, dream house, dream date, American Dream, the 787 Dreamliner, fantasy football, etc. God forbid this mentality leaks into such a serious thing as the diamond-hard science of outer space exploration – because I don't want to be on Mars looking at my time-travelling-clown-faced baby when I could be back on Earth merrily tip-toeing through the nuclear wasteland the Mars advocates had promised.

What's interesting is just how many "scientists" are pulling down a salary that's based purely upon their ability to convince impressionable, ignorant, and gullible benefactors, venture capitalists, and lobbyists, using nothing but optimism and respectability – conferred by one advanced degree or another, published articles, books, etc. – that such things as living on Mars, Goldilocks planets, speed of light travel, and asteroid mining are not only possible, but within arms reach – even though they are just the opposite: impossible, nearly impossible (which for practical purposes is the same as impossible) or could *never* be cost effective or worthwhile, now or ever – and which are all therefore ridiculous. In other words, a complete fraud.

Many examples of legitimate science have turned into junk, and the junk science is incentivized by all the money thrown at it, resulting in a feedback loop that has become a monster, all of which involves countless and ridiculous claims, promises, predictions, and misperceptions about reality and our powers to control it – while at the same time we're increasingly able to realize what we had always only been able to imagine and the border between science and science-fiction becomes a double-edged sword. But for the sake of humanity, the exploration of space needs to be guided by hard-edged sanity, not delusional hopes and dreams (or psychos).

As if the words themselves – science, scientist – are just labels to be applied willy-nilly, conveniently stolen by sociopaths and maniacs, then disguised – a label we sometimes stamp on something to say "this is real smart" even when it might be "real stupid." The connotations of these words – scientist, engineer – have surpassed their denotations, in which we automatically assign a value or worthiness to something that is undeserved. Every astrophysicist isn't an Einstein. More and more things are called "junk science" – this is evidence of an increasing decrepitude.

(I don't mean to be sarcastic, but) why not just terraform a part of Earth to make it more like Mars? Then we can grow gills and toughened skin. This might be a practical, preliminary step toward a Martian future. That way we can avoid the costly effort of moving even a tiny portion of our planet to another – while venting much of the pent up mania.

EVEN IF WATER DID ONCE FLOW ON MARS: THIS DEMONSTRATES how irrational scientists are (and isn't science supposed to be rational?) that they allow themselves to be intrigued that water might have flowed on Mars, because that wouldn't change the present, mitigate the other list of critical problems that make human life – and all other life as we know it – impossible on Mars. In terms of a future human colony or civilization, it doesn't matter what might have been on Mars – Mars' ancestral history is really quite irrelevant – or do scientists suspect a *larger cycle* whereby Mars would some day return to a watery, perhaps even oxygenated planet? Or that there might still be a few tens of thousands of years of some élan vital left in the girl, we just have to coax it out of her. That's the problem with the typical Mars advocate engineer or scientist mentality – such epistemological fallacies are beyond the scope of their awareness, training, and perhaps even comprehension, brainwashed as they are by the weirdly naïve "can do" mentality pervasive in those professions. There is convenient evidence for this: In the 20 years between 1991 and 2011 NASA spent over 20 billion dollars on projects that were cancelled. It was also reported [Elert, popsci.com 9/25/12] that the Department of Defense spent over $46 billion in cancelled weapons programs in the second of those two decades alone. NASA's vision obviously far exceeds its reach. Aren't engineers supposed to be bright? – with NASA engineers among the brightest? Their dreams are handicapping our ability to cure, and diverting attention from, problems here on Earth. Mars advocate optimism is clearly outweighed by the laws of nature which they can't defy try as they might. A lot of these cancelled projects are nothing but high-minded jobs programs for otherwise unemployable engineer/oafs – which in turn may be a reflection against the hiring/recruiting "industry" which, like many things in life, is crazy and out of control – we hire engineers who dream or seem like they'd be cool to work with but who can't fix our problems, make things that work, or devise a cure. We celebrate the NASA engineer with the Mohawk, not the inventor of the breakthrough technology or cure. The recruiting industry has subverted the hiring process – thanks to search engines that read the resumes rather than recruiters – where success is determined more by loading the resume with the right buzzwords than by skills or experience.

There's a problem with the PR machine at JPL – even incidental circumstances turn into "holes in one." Mars is riddled with craters, such that targeting a flat plain – as a landing site – might be a challenge. So if a flat plain was the intended landing site (for the Opportunity rover in 2004), how could the apparent miscalculation be inflated into a "hole in one" after it inadvertently rolled into a crater? (Wikipedia.) When we're sending people to Mars, I would expect the odds to be completely in our favor, not the long odds one associates with "holes in one," which suggests we have a long way to go. If we're going to be sending tens of thousands of human beings to Mars, in dozens if not hundreds of rockets, over the span of a few or many decades, it obviously needs to be with a bit more degree of confidence and precision than would be implied by "a hole in one." So it was actually not reassuring – eight years subsequent to the landing of the Opportunity – to see all the hoopla and backslapping at JPL at the moment the Curiosity rover landed a few years later in 2012 (carried on live TV) – as if they were nothing but a bunch of ragtag dropouts fully expecting failure. Are they not the best engineers in the world? Using state of the art technologies? Can you imagine this behavior the next time your plane landed?

On December 12, 2012, CBS News announced the discovery, by the Curiosity rover, of inexplicable ghosts and shadows. No, I'm kidding, but they did detect a "trace" amount of carbon. Just another NASA taunt? Yes! Here we go again with more "trace amounts." Why does NASA persist with the ridiculous rule that trace amounts of certain chemicals or elements can have wonderful implications? If there's life on Mars, it seems there wouldn't be just a little bit that's barely detectable, or inferred. This was on the evening news and preceded by a few days of preliminary announcements by various other media, so I was looking forward to the full report, but NASA second guessed the discovery by claiming that the trace amount of carbon may have been contamination from the rover, and that, absent this concern, this "first soil sample" test result (from this particular rover since it landed three months previously) was not unexpected. OK, so the big news was that there was...no news.

So – what exactly is the green light NASA is looking for? The entire planet is already mapped out, and has been for years, since before the first rover, Sojourner, landed in 1997, which would have to have been the case in order to have landed even that – and it's all desert. We've been taking satellite pictures of Mars for years. Do they think it will suddenly be different? If they are, in fact, not actually expecting any breakthroughs from the Curiosity rover (according to NASA project scientist John Grotzinger telling reporters on 12/12/12 during the same press conference), then what? Doesn't it seem someone at NASA could admit the obvious? Why do they leave it up to disinterested parties like me to break everyone's heart? Is it an obligation of modern life that I be involuntarily and repeatedly drawn to the edge of my seat, at the expense of more relevant nonsense? NASA has worked itself into a "boy cries wolf" situation. Can they redeem themselves? One thing is certain: the next rover will even more *precisely* determine the complete lack of life than ever before, while more "trace amounts" of random substances continue to play with our perceptions of life as we know it – and the photographs will be even more amazing!

NASA - what a jobs program! It seems they could launch another rover sooner. But it really has nothing to do with "basic science." The facts of Mars are entirely nailed down, with more than sufficient precision. It's just geeks playing with tech in their sumptuous Pasadena studios. In the best playground of them all, right there in sunny southern California. Surfs up – who has time to innovate! Going to Mars no longer has anything to do with a "mission" or other sort of altruism or "national pride." NASA provides neither a service nor a marketable product, and we've reached the point where our technology exceeds our abilities to use it. What ARE those guys up in the ISS doing? – writing "The NASA handbook on zero-gravity sex positions?"

Why do we need to be so reverential about NASA? It seems that of all the sciences, they've done the least toward making our lives better. They have launched a few satellites. Does any of that make my life better?

We seem to have reached the point in science where projects are either just so far out (like CERN) and have nothing to do with human life whatsoever; repetitive, being repeated by separate groups in various forms but ultimately the same thing; or redundant, where budding scientists recreate established science as a learning tool. The CERN project told us something we already suspected, about the existence of the so-called Biggs-Hoson particle, so even that may be as redundant as it is far out. But in space, everything is weightless, even scientific ennui.

29

Science Writers

And everyone is in on it! Like a mass psychosis enveloping our entire human population! A whopper of a maladaptation that's crept into the gene pool.

It's one thing for science-fiction writers to promote the idea of living on Mars – people love fantasy. But when real scientists walk arm in arm with the science-fiction writers, each embracing the others ideas, the line between the two becomes hazy. There's a lot happening in that hazy area right now, including claims that such sci-fi tropes as "warp drive" may soon be possible. Other examples of the "science writer" out of control on a happy-go-lucky bender are numerous, found in pop-sci magazines and websites, who ignore, deny, and trivialize not only basic science but ethical, political, economic, biological, and cultural obstacles in their wild predictions and promises concerning such far out topics as speed of light travel, alien visitations, worm holes, and black holes. Readers lap it up, it trickles into the mainstream as "scientific truth" and even the scientists get brainwashed by it. But there needs to be a diligent and skeptical analysis of exactly what the Mars advocates are proposing regarding a full-scale settlement on Mars, including exactly why their plans aren't just unreasonable now, but most likely forever.

For example an article at Universe Today (Internet, November 2012) described how the rover Curiosity was taking the "first ever" radiation measurements from the surface of another planet. I won't reprint the article here, but it contained so much misinformation – propaganda – that I'm surprised it got published. For one, it's not true that Curiosity (the *fourth* rover) took the first ever radiation measurements from the surface of another planet – we already knew, and had for years, that there is a huge amount of radiation on the surface of Mars that is completely unsafe for humans. Our ability to live on Mars will not be achieved by successively narrowing down the precision of existing measurements. It's disingenuous to say that something is the "first ever" (with the attendant excitement and gee whiz hoo haw) when it simply isn't. The article should have used the known dangerous levels of radiation as the premise to explain why man can't live on Mars because of this, instead it makes the opposite conclusion! Well, what do you expect from the uncritical, slaphappy Mars advocates? Thanks, Universe Today, for contributing to the National Averaging Down Project now underway regarding our global "collective intelligence."

Another example of the warped, Mars advocate journalism is the excited report in the journal Science (Sciencemag.org, December 2013) of a 3.5 billion year old dry lake bed on Mars – as if this was an invitation to a neighbor's pool party. Would this be part of a 3.5 billion year cycle, such that a human colony on Mars would be perfectly timed? Might our current state of technology and a 3.5 billion year cycle on Mars converge? Of course not! It's just those *science writers* again. The truth is this just shows how dead Mars really is. But I would expect that lakes and other "Earth-like" phenomenon would occur on *many* other planets over the course of cosmic time – and I would expect there to be even younger ones on Mars. The problem is these things don't make Mars any more livable *today*.

Some scientists suggest that manned missions to Mars won't be possible for another 20 years or so. Certainly this must mean that we, the literal we, aren't going to Mars. Oh shucks. But they don't want to lose their jobs so they substitute manned with unmanned "research projects" – with virtuous and self-aggrandizing names like "Endeavor" or "Mars 500" – and many others wild and crazy that are eventually cancelled. But that twenty year margin gets them off the hook, while still

seeming "soon enough" to be within reach – because humans operate
on the dichotomy of soon or never. We don't want to completely let
go of the dream, for it to be killed off – but we seem to have passed
the rubicon so this dream may be impossible to let go, where the only
possible outcomes lie between pointless and painful.

MASS PSYCHOSIS

Why do science writers feel obligated to be part of the propaganda?
And everyone is in on it! It's almost as if there's a mass psychosis enveloping our entire human population! A whopper of a maladaptation
that's crept into the gene pool. As a writer, my feeling is "these guys
are crazy! Why would anyone want to go to Mars? I hope they realize
how foolish they're being, god forbid this continues." I can't help but
think that's how many science writers feel, at some level, toiling away in
their strange esprit de self-censorship, working more to maintain some
illusion than to inform. In their happy-go-lucky articles they say things
like "I don't know how they'll do it, but I'm sure they will, I hope they
do," or similarly trite sentiments, which seem to be hypocritical and
a big lie. The Big Lie. Because they would never actually go to Mars
themselves, while at the same time wish this for others. Or is that the
standard now, for journalists to be hypocrites? There has never been
anything so deserving of skepticism – living on Mars – yet science writers seem never to express even the slightest bit of it, or just a bit but then
veer back to the dumb – automatic – sure, OK, sounds great when are
we going? This is, in fact, the nature of propaganda. Since few people
doubt the science writers, disinformation (regarding the prospects of
humans living on Mars) gets repeated by journalists as well as parents
and others in authority, thus causing the disinformation item to become
a "well known fact" while no one repeating the myth is able to point to an
authoritative source. The disinformation is then recycled in the media
and in the educational system, resistant to intervention or criticism.

This attitude, taken by even the most legitimate science writers regarding the exploration of space is, presumably, entertainment driven,
but otherwise provides clear evidence of brain washing/conditioned

thinking/automatic thinking – or has the effect of doing that – while merging fact and fantasy. Tolerating this for the sake of amusement or selling newspapers or magazines is one thing, but at the point where people think they can actually live on Mars – and sign up for training! – and build rockets! – one needs to confront the matter head on, free of the space cadet mindset.

SCIENCE WRITER AND TEACHER OF HUMAN EVOLUTION (PORTLAND State University) Cameron Smith, writing in Scientific American magazine in January 2013, suggested that woman about to give birth "get up to an orbiting space station," which would have an artificial gravity greater than on Mars' surface. This is typical of the many "solutions" proposed by engineers or others in the space industry – and literally bursts free of any attempt to be reasonable. It is a special refinement of ridiculous. And like the spacesuit, it is a Procrustean bed. What exactly is Mr. Smith thinking – that he's writing for Mad magazine? Looks like we have another "specialist!" – even better, this is the *Mars advocate* specialist – consistent with Mars advocate logic, but here you have it in all its glory. Because if Mars isn't conducive to human reproduction – or any aspect thereof – THERE CAN'T BE A COLONY ON MARS! If you have to LEAVE THE PLANET for something as mundane as giving birth – you may as well call it quits right now! *What aspect of life on Earth requires that we...leave the planet? Let me think hard. None.* Life is pretty simple, and most of it is actually quite free of technology, *especially* things that involve reproduction. Nature has kept this *real simple.* You don't want your *beloved engineers* to add any degrees of difficulty to this, especially if you're already way outside of your *comfort zone.* Right now, women actually don't need to go anywhere to give birth. Sometimes they go to a hospital for the sake of expediency. But on Mars they'll have to *leave the planet*? Mr. Smith has no idea how dire the implications are of his suggestion! How can none of this have occurred to him? He's so trapped in his Mars advocate brain! First, it's not the tug of gravity that causes child birth – it's hormones, combined with muscular contractions, that expel the baby. Second, as if "getting to an orbiting space station" is the equivalent of riding a taxi through town! – and this should be the norm on Mars? Third, it would make

more sense to just make c-sections the norm on Mars, or to create whatever environment is most conducive to childbirth *right on the surface of
Mars,* even if that meant making a giant, (slowly) spinning centrifuge, in
keeping with Mr. Smith's odd vision.

Who are these science writers? Their writing may be grammatically
correct, and they may have respectable credentials – but is there no
limit to how crazy an idea can be before someone challenges it? And
the respectability of the publication suggests that Mr. Smith's idea was
not made in jest. Between the original thought and its publication – I
don't see how his idea can be defended. Is there no equivalency between
present "natural processes" and future? One might expect his idea to be
a misogynistic joke (but which clearly it isn't) or written by a C or D level
high-school student on a paper handed in late, where the student was
scrambling to write anything to get it finished. Or maybe Mr. Smith gets
his ideas from C or D level high-school kids. *Or maybe he was drunk
or drugged out!* In any case, his specialty is clearly science-fiction and
not science. So, why is the bar set so low for science writers today, even
among the most respectable magazines – as if we were still in the 18th
century or earlier. *No – it's because this is how scientists and engineers
think!* They just don't quite get the whole picture, they're all so focused
in their niche, with their *acquired idiot savantism.* Or they reach a level
of incompetence, move into writing and, still somehow hanging on to
whatever respectability they might have had, get published! And the
pile of junk science just grows – some day it may reach Mars!

Let's pursue the implication that I referred to above (that leaving
the planet should become usual and ordinary) made in Mr. Smith's
suggestion (that pregnant women should go to an orbiting space station
to give birth) and that is far-reaching. If a creature living on Mars –
such as someone born on Mars in an imagined colony – were to move
to Earth, that doesn't mean they're moving to the ISS (the space station
orbiting Earth). Or to the Moon. It means they're moving to somewhere on the surface of the Earth, and which does not suggest side trips
to either of these two other places, because that would suggest a very
weird but constant state of disruption and transition. There's a certain

"groundedness" that is associated with being healthy and fit, in having a *normal life*, which takes on a literal meaning in this new context. I would expect *moving to Mars* to have the same general meaning. That this person would be living on...the surface of Mars. Not in a lab orbiting Mars, or anywhere else! Nor would there be an ongoing need to inter- mittently *leave Mars*! What else can one expect but that a normal life on Mars would have some equivalency to a normal life on Earth – obvi- ously – and normal life on Earth does not require that we intermittently leave the planet – quite the opposite, because life, by its very nature, is confined to a complicated set of very particular circumstances which, generally speaking, *precludes* any possibility whatsoever of *leaving the planet*. So in this regard, the idea that one would need to "leave the planet" for something so ordinary as child birth becomes particularly outrageous. Plus there would be weird legal matters concerning "place of birth" and citizenship status, with all kinds of ambiguities that would never be settled.

It's amazing how modern and sophisticated the Mars advocates imagine life will be for their colony! Satellites don't stay in orbit for- ever. They break down, require maintenance, spin out and crash land – things we take for granted and, for the most part, don't care about. We just launch another one – and we have more than ample means to do so – or launch something *better*. Does Mr. Smith expect Mars to have its own space industry, such as would be required for the repair and maintenance of a birthing facility orbiting Mars? What would it take to get to that advanced stage – and how would babies be born in the meantime? A Martian colony could never survive if it depended on such a Kafkaesque arrangement. These and other conundrums litter the Mars advocate landscape.

30

Microgravity

Elon Musk is the ISS garbage man.

THE CREW ABOARD THE ISS CREATES TRASH AND NEEDS GROCER-
ies. Mr. Musk – via his SpaceX Dragon and Falcon spacecraft, which are
unmanned – delivers the groceries and takes out the trash. Mr. Musk
is the ISS garbage man. So, are you in shock? In terms of technical
achievements that get us closer to Mars, this does not even register. Do
you still think Mr. Musk has anything to say about "colonizing Mars?"
Would you take advice on moving to Mars from the guy who runs the
lunch cart in front of your office building? Or does the Ford Motor Company (or whoever makes your car) have a say in where you live or travel?
Is our inner space garbage man guru qualified to – singlehandedly –
take control of destiny's child?

> **Junkyard Mars, garbage man Elon, there you have it
> in its full depth and breadth. That's the breakdown.
> And will inner space trash collector lead to outer
> space gold-mining?**

So, SpaceX is like a lunch cart for 6 people floating about 250 miles
above the Earth. Put in this light, is this really deserving of Smithsonian
Magazine's 2012 American Ingenuity Award for Technology, given to
Mr. Musk? Or is Smithsonian Magazine simply overcompensating him
for calling him a weirdo in a December 2012 interview? This signifies,

if nothing else, that the Mars advocate fever has boiled over onto the streets.

Right now there are millions of people starving to death worldwide. Over twenty-thousand will die today. Today. What would be the prize to send a rocket to Africa? Oh – screech...I didn't mean to get preachy and practical. Anyway...

It must have something to do with the ISS, which begs the question: what is the ISS? It's a lab in Earth orbit staffed by a crew of 6 people (crew members are swapped out every 3 months or so) – which doesn't seem like a lot. How much science can 6 people do? Did you hear me? I said HOW MUCH SCIENCE CAN 6 PEOPLE DO? They study micro-gravity, among other things, but they're actually the subject – the guinea pigs – of a larger experiment: to see if and how man can live in space – but being within arms reach of NASA affords them exquisite advantages, and it's simple-minded to think that life on the ISS can be extrapolated to living on Mars.

FROM THE ISS TO MARS?

And why would you WANT to extrapolate from the ISS to Mars? Whatever aspects of life on the ISS that can be extrapolated to Mars – is this a reason to celebrate? Certainly the key features of the ISS are the following: 1) It is in constant contact with ground control, allowing all parties to communicate visually and verbally – and of course this is all done *seamlessly*. 2) It can be serviced by either SpaceX or the Russian space industry (which has two rockets docked at the ISS at all times). These services include the delivery of food, disposal of trash and gar-bage, and repair and maintenance of the ISS and 3) these services can be provided in a timely manner, in a manner that might be considered convenient. All these things are essential to the operation of the ISS *in a way that can't be overstated!* One could say these things are the heartbeat of the ISS – but none of which can be extrapolated to Mars! A colony on Mars will have none of these things! In the same way we take our own heartbeat for granted, so we take these aspects of life on the ISS for granted.

Or perhaps there is some aspect of living on the ISS that I can incorporate into my life on Earth? Besides everything being upside down. The ISS is a ghastly, weird experiment.

All I want for Christmas is...six months on the ISS? Tell me, what are the best things about living on the ISS? It seems there is NOTHING about the ISS that can be extrapolated to Mars. Because everything that should be (the three features listed above) CAN'T BE, and everything that should NOT be (food in a tube, adverse affects of microgravity on physiology, and other hardships) WOULD BE. The modern expression for this is LOSE-LOSE. And we're supposed to be celebrating all this?

Man evolved to live in full gravity. Do NASA and other scientists think that there wouldn't be adverse consequences to living in anything less than full gravity? That our physiology might somehow be enhanced? Maybe they think there's a "microgravity gene" we can just switch on or off, as if we evolved in anticipation of living on Mars – wouldn't that make us the "smart" species! Of course, that must be it! Maybe that skin tag is really a microgravity switch! – and if I just scratch it off, I'll be in full Mars mode!

MICROGRAVITY? IS MICROGRAVITY KEEPING ANYONE'S GRANDmother alive, anyone's newborn baby alive? No – so what's microgravity and is it anything you should care about? Gravity decreases as you get further from Earth – hence microgravity – and Mars has less gravity than Earth.[1] On the ISS there is no gravity, unless it's artificially created through spinning. Things behave differently in microgravity. In ways we can't expect or even imagine. It's the next big thing if you're a believer, a Mars advocate.

So is the ISS keeping millions of people employed? Certainly not. Tens of thousands? Many people are employed, directly or indirectly, in the name of the ISS, it is, after all, international. This may be the best thing that can be said of it, since the value of any research done so far is tenuous and very long term and very likely *will never be realized.*

....

1   Microgravity is zero-gravity. Since the gravity on Mars is 1/3 that of Earth, Mars has reduced gravity rather than zero-gravity. Reduced gravity causes a loss of muscle and bone mass proportional to the reduction in gravity.

The ISS is no longer serving any purpose.

Microgravity research only has relevance for living on Mars and only a Mars advocate would advocate the continuation of the ISS – unless someone plans to manufacture massive amounts of chemical compounds that can only be made in microgravity – because only with massive amounts would any manufacturing process be profitable – or can foresee this potential – which is implausible because the science that we know is established in Earth's full gravity, and a parallel science based on microgravity might well be considered provocative, even tantalizing, but ultimately useless – BECAUSE WE EXIST IN FULL GRAVITY! – and the ISS is not equipped for the manufacture of massive amounts of chemical compounds and an expansion of the ISS for this purpose would be cost prohibitive – then the ISS is no longer serving any purpose – other than frivolous science for its own sake but certainly not pertaining to matters of national security, as stipulated by NASA policy and the other countries involved. The international aspect of the project itself may serve as a form of détente, in some token way – as if the fact that we're all already riding this freight train isn't enough.

And if there were any potential to a microgravity manufacturing environment, it makes infinitely more sense to devise the means to do this down here on the ground. We can make enormous microgravity factories which employ tens of thousands!

Just tap into some of that good old American hubris ridden engineering potential that for some reason is being reserved for outer space and other planets! It's hard to imagine that anything being done on the ISS will serve to remedy any of the inequities or iniquities down here in full gravity, because that has so far not been the case.

Since the ISS first went into operation as a manned orbiting space station in 2000, one might have expected it to grow in size to accommodate a larger and larger crew. But it hasn't. What does this tell you? And the living quarters for the six crew members are extremely cramped. One might even have expected that, by now, it would have grown into an enormous off-world colony, for the wealthy or other elite – the adventurous! – like the ring-shaped orbiting habitat in the movie

Elysium – which some think is plausible but which demonstrates – and so beautifully – how different the realities are from our expectations, even where the expectations seem simple, but especially when these expectations require us to *defy gravity.*

Smartphones work like magic. Rest assured that anything that is magic on Earth will not also be magic on Mars.

Of course, existentially, everything that works, works BECAUSE of gravity and air pressure and other laws of Earth-nature that would not be on Mars. It would be an outrage to think that our science dollars – or any dollars – would be diverted to develop a parallel science in microgravity for the sake of a few dozen, hundreds, or even thousands of people on Mars, since everything would have to be reinvented for use on Mars. For example a 3D-printer (that works on Earth) wouldn't work on Mars – it would need to be completely reinvented. 3D-printing is actually a convergence of several technologies – which, like the smartphone, work together magically. But rest assured that anything that is magic on Earth will not also be magic on Mars. The four rovers are engineering marvels, and required the invention of designs uniquely suited to the Martian environment. They were *designed* to work on Mars. But reinventing everything on Earth is highly improbable (since we don't have microgravity based R&D or other Martian laws of nature) and what's highly improbable on Earth will be impossible on Mars, which would completely lack R&D altogether – since that would rely on Mars science but there is no Mars science! Mars science itself would have to be invented! – and this could only be done on Mars. It's a proverbial Catch-22.

The 3D-printer is simply one of many examples that demonstrate the far-reaching limitations regarding our expectations of quality of life on Mars. One giant step that would enormously benefit the Mars advocates would be to develop a giant-sized microgravity R&D empire, on Earth, to reinvent everything for Mars – on Earth of course because such an enterprise would need to be vast – global – and which could never be scalable to Mars.

Earth itself is, in a sense, an R&D empire, but as an aggregate effect of millennia.

So it stands to reason that a Mars R&D can't just plop into existence – but without this, a transition to a Mars with earthly standards would be impossible. A giant-sized microgravity R&D empire on Earth? You heard it here first.

WHO WOULD HAVE EVER THOUGHT OF MICROGRAVITY AS SOMEthing that could possibly be useful for research – that's the real kicker! Of all things! Micro-gravity? After all gravity isn't exactly something that can be manipulated – it's what scientists call a constant, a law of nature.[2] The whole thing seems not only unintuitive, but counter-intuitive, to think that, since everything that we know occurs on the ground, in full gravity – and always has – that gravity might actually be...inconvenient. Oh gravity, release me from thy mortal grasp! It's got to be another one of those "mad scientist" things! Man has always dreamed of flying, and thanks to the Wright brothers, we can. But, intuitively, since birds can fly, there's no reason aerodynamic laws would be different for us – and they aren't. There's therefore nothing magical or foolish about it, the fact that we can fly. But microgravity? Microgravity isn't the reason we're exploring space. It's not even a fringe benefit but rather a queasy, very undesirable side-effect. It's the thing that causes muscle atrophy and bone loss. Microgravity be damned. It is anathema to NASA and the international space industry while at the same time the ISS crew plays with it and has the gall to call it science! It's science like when you throw litter out the car window at 100 miles an hour is science. Playing, even where the laws of nature are different from down here on the ground, is not science. It's playing. So it's kind of a stretch to say that

2 At least we treat it like a constant – gravity actually varies from place to place to place, in miniscule amounts, such that your weight would be a tiny bit different at the top of a mountain than at the bottom, but these differences are generally, and historically, ignored – even though they've recently mapped differences in surface gravity on Mars, so they can more accurately pinpoint target landing sites for craft we send there – but this may just end up overcomplicating things, and add to the pile of information overload, or "make-work" for another poor MIT grad, since it seems any such fine-tuning of target landing sites would be cancelled out by the larger circumstances known to exist – and the random ones like windy weather – something Mars has in common with Earth.

we're "studying microgravity" on the ISS and to act as if it's a valuable objective – or that such a thing is or may be critical to our survival down here on the ground.

Certainly no one envisions a microgravity manufacturing facility in space, with some kind of shuttle service, or "space elevator," providing a steady supply of specialty products to Earth – how ridiculous! On the hierarchy of ridiculous, nothing ranks higher than the "space elevator" – perhaps not even a Martian colony. It deserves its own chapter (in a different book).

We HAVE TO STOP THINKING THAT ALL THIS WOULD BE POSSIBLE if not for just one more threshold that needs to be crossed, where safety is no longer an issue, as if travelling to and from space could suddenly, or not suddenly, become risk free and ordinary. As if travelling back and forth between Earth and satellites could ever be safe! That point will never be reached. Every docking of the Shuttle, then SpaceX's Dragon, with the ISS is considered a miracle! – according to those involved, dazed by their own success. Just look at the Space Shuttle program, where each of five Shuttles was initially intended to make 50 flights per year, but ended up making an average of only 4 per year – it doesn't matter why! – it was retired! – it's history! Look at the SST airliner that was retired due to safety issues – it's history! Wherever you have people in things that can crash, our tolerances for safety transcend anything else, and thanks to gravity, it's inevitable that things will continue to crash.

> As man presumes to defy gravity, and other laws of nature, things don't get less dangerous, they get more dangerous, because we're always raising the bar, incorporating ever more complex features into our designs, so the odds become less and less, not better, that we could ever cross the line into risk free and ordinary space travel.

The high rate of success of the SpaceX rockets that docked with the ISS (at least up to 2013) is due in large part to the rockets being unmanned, and therefore free of the complexity imposed by design issues otherwise necessary to support and protect human life. But watch what happens when SpaceX raises the bar, and complicates their beautiful design with all the amenities needed to keep people alive – comfortable! – in the vacuum as they hurtle through the void. As things get real complicated, we're once again up against that impenetrable boundary – with "safe and ordinary" just on the other side – and the story will become, like the Shuttle and SST, tearful and tragic.

It's not as if we're getting any less frail as the bar is raised into more and more extreme and unnatural and nature defying territory. Quite the opposite – we've become more frail, we humans, and sensitive and fine-tuned to our world. Each human generation is arguably more adapted to Earth than the previous – which means we're more fragile. In addition to this, we're preserving more and more inferior traits (premature births, birth defects, etc.), qualities that in nature would be allowed to die off.

Granted that control of "inner space" (the imaginary sphere that contains, for the sake of this discussion, Earth and the Moon) is, arguably, crucial for modern technologies and survival on Earth (forecasting the weather, studying ozone and global warming, and GPS navigation are a good chunk of the iceberg) but this doesn't require that people live in space or have anything to do with the exploration of outer space or other planets. Satellites can be put in orbit and operate without the need for manned spacecraft – the conquest of inner space has relied only partly, fortunately, on manned space flight. And look how successful the four Mars rover projects have been, all completely free of manned space flight, in addition to extensive satellite reconnaissance of Mars – all due in large part *because* they're unmanned. One might, then, consider this – unmanned – exploration to be the pinnacle.

SpaceX is tiptoeing within the safe zone of inner space, as well it should, but pushing its limits in the way it has promised (carrying people to Mars) isn't cool crazy or sexy crazy, it's deranged crazy.

MORE ON THE ISS

Space Shuttle to the rescue! Soyuz and Progress to the rescue! – which are Russian spaceships (the manned Soyuz and unmanned Progress). At least one Soyuz is docked to the ISS at all times as an escape vehicle, but to be replaced by the PPTS, also Russian.

Mission Control to the rescue! Such rescues will not be possible on Mars!!! The ISS may at least be good for studying the long term reliability of life support systems – but which are, evidently, not all that promising if one's objective is fail-safe machinery. If something breaks or malfunctions on Mars, there won't be someone coming in a few days or weeks to the rescue! Put this book down right now and think about this.

In 2005 ISS personnel tapped into the oxygen supply of the recently-arrived Progress resupply ship when Elektron oxygen generators were "plagued with problems" – which may be its own never-ending story.[3]

Repeated failures of an oxygen generator on the ISS. Will this extrapolate to Mars?

3   Tariq Malik (4 January 2005). "Repaired Oxygen Generator Fails Again Aboard ISS." Space.com.http://www.space.com/missionlaunches/exp10_elektron_050104.html.

31

Reproduction

Will we love our Martian babies? Will we dream of Earth's blue skies?

The first babies born on Mars might look deformed, hideous, like time-traveling clowns caught in the wrong time – a new age of freaks and geeks!

NEW RESEARCH ON MAMMALIAN REPRODUCTION, BASED ON EX-periments with mice embryos in 2009, has shown that a lack of gravity interferes with an egg's ability to properly develop (http://www.tech-novelgy.com/ct/Science-Fiction-News.asp?NewsNum=2670). Add to that the markedly higher doses of radiation an astronaut is typically exposed to or would be found on Mars.

- It has been known for decades that some animals fail at pregnancy while aboard orbiting spacecraft.

- Lower atmospheric pressure, and oxygen levels raised to compensate for this, both interfere in vertebrate embryo development. (Scientific American, Jan 2013.)

- Experiments simulating zero-gravity conditions detail biological difficulties in mammalian reproduction and development in space. Teruhiko Wakayama and his colleagues at the RIKEN Center for Developmental Biology in Kobe used a 3D clinostat to conduct in vitro fertilization (IVF) experiments with mouse sperm and ova in a simulated microgravity environment. Microgravity led to an overall reduction in the rate of blastocyst formation after 96 hours of culture, and closer examination of these blastocysts revealed that the differentiation of embryonic cells into trophectoderm—the tissue that nourishes the embryo and ultimately contributes to placenta formation—was markedly impaired.

- As mentioned in the chapter "You Like to Breathe, Don't You?" the Mars gravity biosatellite experiment was cancelled due to lack of funds.

- There is abundant evidence that microgravity affects virtually every aspect of plant growth. This means we can't just throw seeds down on Mars like we can on Earth.

- Are you in shock?

All this strongly suggests humans can't live in zero or reduced gravity – certainly not in terms of living on Mars as a robust, thriving, self-sustaining colony – that humans can't "go forth and multiply" on Mars.

32

Sex in Space

‖‖

**Scientists classify us as "animals," so let's not
pretend that man in space will be less an
animal than man on Earth.**

YOU'D THINK THEY'D BE HAVING SEX LIKE CRAZY ON THE ISS, TO
see if fertilization and embryo and fetal development can occur in microgravity – something which could be done on the ISS – but which
is contrary to ISS policies, which is ironic because we would obviously
need to know about this in order to live on Mars. The ISS seems perfectly suited to this, it is, after all, a microgravity lab with six people living
on it. If you want to see how something works in microgravity, the ISS
is the place to do it – it is the *only* place to do it. If one had never heard
of the ISS, but that Mars had reduced gravity and they were planning on
a colony, they might suggest something like the ISS to study human reproduction in microgravity – and then be shocked that there was already
such a thing but that it wasn't being used to study human reproduction.
But it's the truth. Or is it?

This subject – with the greatest questions of all, ones that can be
easily answered – is being avoided! This prudishness may be one of our
biggest obstacles in getting to Mars. Well, as a jobs program, designed
to keep the greatest number of NASA engineers working the greatest
amount of time, maybe they don't want their gravest suspicious confirmed: that human reproductive biology doesn't work in microgravity
nor therefore, it could be reasoned, in Mars' reduced gravity. Either way,

according to one account, by a crewmember of the ISS, "we can't have sex on the ISS because it might make the ISS structure jerk around." Well, someone's got to loosen up at NASA, because at some point, and before anyone sets off for Mars, we have to know what the laws of attraction are in space, and to that end, people (whether under the guise of astronaut or scientist) need to be allowed to express their basic instincts in all their glory – if we expect a true human colony to survive on Mars.

Without realistic experiments, the ISS has become high-minded pointlessness. The ISS exists primarily to benefit SpaceX, it seems, and SpaceX exists primarily to benefit the ISS (and perhaps a few other inner-space projects) – a strange symbiosis with no immediate benefit and whose long term benefits are questionable at best. But wait – if nothing else it's a thrilling experience for astronauts who stay on the ISS! And for them to eat delicious reconstituted food, and other things... and bodies that disintegrate! And no sex!

THE HUMAN SEXUAL AGENDA

Unlike rocks or other inanimate objects, living things procreate. As living things, we exist to procreate, which means we exist to have sex. To civilized man, sex is merely a component of life but, more existentially, it is life itself. NASA and the rest of the space industry is very uptight about this, and seems to treat humans, and along with it the human condition, as if we were rocks, inanimate, not alive. The space industry has put a very G-rated, family oriented face on the space program, but let's put down our guard a little and not be too uptight about it. Because Man, in fact, does not exist to "do science" or "perform experiments." Scientists classify us as "animals," so let's not pretend that man in space will be less an animal than man on Earth. After all, we are as much in space now as if we were on Mars – there's no difference! So it's ridiculous for the space industry to maintain this vision of man in space as being purely scientific or task oriented. Our mission "in space" will be exactly the same as our mission on Earth: to live, and, in keeping with the same sense of adventure that is so madly promoted by the space industry, and given that we are sexual beings – first and foremost – this means that the human sexual agenda will be living large – AND SHOULD BE – as we venture forth. As if such a thing could be stopped. Or repressed. Do you allow the government or commercial enterprise to control your sex life now? Then why would you allow this anywhere or anyplace else? But if that's the case, we should assume that they are, in fact, having sex on the ISS – how could they NOT be? There are, after all, both men and women on the ISS.

> **The ISS may, in fact – at this point – be nothing but a high-flying sex club or brothel in space! Everybody high five! Because if they're NOT having sex on the ISS, in accordance with policy, the implication is far-reaching: that there is really no intention of colonizing Mars!**
>
> **And if they're not having sex on the ISS but for reasons NOT related to policy, then clearly there is no sexual desire in microgravity and a colony on Mars would therefore be impossible.**

Will human sexuality be able to flourish on Mars, as it does on Earth? The idea of "usual and customary" may take a new direction in space, or have no meaning whatsoever.

> **But one thing is certain: the human sexual agenda must be fulfilled. If this is not possible on Mars, a colony on Mars will be impossible. In this regard there is no telling what pleasures – or horrors – await us.**

MORE ON SEX

It will be interesting to see how human sexuality develops on Mars. Will Darwin rule, or selective breeding? On Earth, monogamy is the standard – will this carry over to Mars, or will hedonistic "adventures" be the "new normal," the mission does, after all, cater to the more adventurous among us! Will Club Mars be the new Club Med? Will I be able to shuck and jive my groove thing? Will there be hotties on Mars?

There'll be no Christie Brinkley's born on Mars. I shudder to think what a "pin up" would look like on Mars – perhaps a cross between Cher, an aardvark, and an iguana – HOT! Media exported to Mars may have to be censored to exclude Earth-centric standards of beauty that, on Mars, may be considered repulsive. What will the standards be regarding recreational vs. procreational sex – defined by a committee on Earth? – or will things become more freestyle – going "where no man has gone before?" Can we assume man will even have a sex drive on Mars?

We do know that sex is banned on the International Space Station (ISS) – and we learned *nothing* from the Russian Mars 500 experiment regarding this, which ended in November 2011 and in which six men were locked up for 520 days in a room in a Russian research building designed to replicate the isolation that would be experienced in a spacecraft bound for Mars and in living on Mars, but where aspects of human sexuality were not studied – there we might have learned something useful for life on Mars – like do men in space, deprived of everything but each other, go gay, as we know men sometimes do in prison. It's interesting that women were excluded from the study group (of which there were several over a few years) – to cancel out the effects of sexual tension – but why would you want to cancel out the effects of sexual tension? When will these experiments get realistic?

It's (still) uncertain if in a reduced gravity environment, such as on Mars, whether man can even have a proper erection *(space.about.com, John P. Millis Ph.D.)*.

Preliminary tests on mice regarding mammalian reproduction in microgravity are negative (Wakayama), but even if human life can be forced to happen artificially, that wouldn't matter if it can't happen naturally as well, with the usual rock and roll, like on Earth. This is, of course, equally true for all the other species that we expect to bring along with us, plant or animal. An ecosystem could hardly be called self-sufficient if artificial insemination, or some other contrivance, becomes the rule.

THERE'S A BIGGER MEANING TO "THE FACTS OF LIFE" THAT ASTRO-physicists and rocket scientists and entrepreneurs are missing, and I don't mean philosophical, since it's preposterous to think we can redesign ourselves, or be redesigned, and that this will "just work" on Mars – or design an ecosystem here on Earth and think it will "just work" on Mars – or to think that the particular order and arrangement of things in and around us – the facts of life – is a trivial coincidence that can be rearranged to suit our convenience or cliché, atavistic *vision*.

33

Mars 500

It seems we're not learning a lot from these deeply involved experiments that we didn't know already.

WE SEEM TO HAVE LEARNED LITTLE FROM THE RUSSIAN MARS 500 experiment, designed to replicate the isolation that would be experienced in a spacecraft bound for Mars and in living on Mars – but it seems isolation would be the least of the numerous adversities imposed by Mars – none of which were part of this experiment! As the Mars 500 experiment was ending, did the men want to stay inside the "sealed room" and not "return to Earth?" Did they refuse to come out of their veritable "man cave?" Because that's what a colony of humans on Mars would need to be – people who want to be on Mars MORE than they want to be on Earth! People who PREFER Mars to Earth! Did the Mars 500 experiment at least establish that? Of course not, it established the opposite: all the men were more than eager to end their isolation – even though the simulation conditions were quite Earth-like – suggesting that even people supposedly interested in living on Mars have a clear preference for living on Earth, which is of course NOT the right attitude for anyone intending to colonize Mars.

An experiment like the Mars 500 would be inherently flawed because NO circumstances that exist on Mars were present! This is what's called *junk science* and was just another jobs program and an example of busywork, which allows us to comfort ourselves by relearning what we already know. Six men were locked in a large room – for 520 days – that was furnished and decorated like a college dorm room, the idea being to see what happens to men in isolation. But if you're hanging out in what seems like a college dorm room with a bunch of your buddies, you're not isolated just because the door is locked and the delicious food is delivered to you rather than you going out to get the delicious food! This is not isolation! It's actually pretty close to how many people actually live! Any hazards and risks to life were first carefully eliminated, and the circumstances *idealized* in order to guarantee the desired outcome – as is true for a lot of science. The men knew they would be safe the entire time – there was no reason they wouldn't be! – that their health and safety were being closely monitored, that they could end the experiment whenever they wanted – like if they got bored – and during which time they were completely free from worries about a reliable and healthy food supply, the air quality, subfreezing temps, or leaks in their highly pressurized living quarters – as they would be on Mars because those would be the circumstances on Mars! The experiment would have been stopped – i.e., the door unlocked – if things "went bad." And the volunteer test subjects, wannabe astronauts, had full knowledge of all this. And of course everything turned out just fine by the time the experiment ended! And of course we learned nothing – other than the fact that some of the men avoided exercising or had trouble with their circadian rhythms (had trouble getting up and going to bed). People avoid exercise? This would actually be a serious problem both in flight and on Mars since we know people in reduced gravity have to exercise daily to offset muscle and bone loss. But of course the Mars advocates think all this proves that man can live on Mars!

It seems we're not learning a lot from these deeply involved experiments that we didn't know already. We already knew that people sometimes get a little depressed and hypokinetic in isolation and that simple diversions like music and exercise are effective remedies. We... knew that already.

34

Star Trek Adventure

Earth is and will always be magnificent compared to Mars – any past, present or future Mars.

THE AGE OF EXPLORATION IS ACTUALLY OVER, HAVING ENDED way back in the 17th century. This is according to National Geographic and all other sources, which had primarily been driven by, among other things, the search for trade routes, although many other discoveries continued to be made up through the 20th century to a much lesser degree.

Anthropologist (and "science writers") routinely attribute our success as a species to the sense of adventure and exploration of our ancestors throughout this period, and the ages leading up to it, as they spread out and established themselves around the globe, but even this seems to be at least somewhat of an exaggeration, with human activity being compressed in retrospect, and that it was probably more of a "wandering" process, tentative in nature, which of course doesn't sound exciting or brave, but nonetheless. Funny how our ancestors are always being attributed with "bravery" and "fearlessness," did they ever just have normal days? Have our creative writing instructors gone too far in their unending insistence to use powerful action words? It seems that the course of history would not always, as we like to think, involve power and drama.

In terms of knowledge for its own sake, is it possible we've already gone too far? Ultimately, the belief that we need to conquer the stars, to distribute our seed and hence perpetuate ourselves – seems like a denial of our mortality, because we are all going to die, and then our species. Alternatively, for those inclined to bizarre fantasy, the idea that a human being could step into a Star Trek existence, and literally travel from planet to planet, or that that would be a reasonable ambition of an evolved species, seems ridiculous, and not just because we lack the means to do so – but because there is no terrestrial analog to this, with groups of scientists driving from town to town – vagabonds – "observing and learning" for its own sake. This has never been the case even where such a thing is possible. We have the means – right now – to live a vagabond existence – a scaled down version of Star Trek, or something equivalent, albeit confined to Earth – to tour around and give help where it might be needed – but clearly it's not the nature of complex species, like ourselves, to be always "on the move" – true nomads – for altruism or any other reason – rather, it is the opposite. And distances between planets in different solar systems, as would occur with people living a Star Trek existence, are really so great they could never be traversed in an entire lifetime – even travelling at the speed of light!

Despite the claim that "it's our nature to explore," that really isn't human nature at all, especially where it takes us completely out of our element. Otherwise we'd already have domed colonies at the bottom of the sea, or bustling cities in Antarctica or the Sahara – we don't and no one wants that – any of which would be easy compared to a colony on Mars! – so why don't we do these other things? And what do the Mars advocates expect to do but turn Mars into another Sahara – a subfreezing cold Sahara?

Tour around and give help where it might be needed? Is this like *sticking your nose where it doesn't belong?* Is that the objective of a Star Trek existence? Imagine star trekkers from another galaxy visiting trouble spots on our own planet. Would we embrace them with gratitude, or treat them as invaders and blast them out of the sky? Even if their objective was to establish peace, would their means for establishing peace be different from our own? Even if they ended up wiping out half the world's population...might this be for the greater good and would you thank them – or build a pyramid to help them back again later?

A writer in National Geographic (January 2013) said "We love the idea, if only from an armchair, of cutting loose from the comforts of everyday life and venturing into uncharted worlds, with no certain destination and no guarantee of safe return." This idea must be peculiar to National Geographic writers because everyone I know is just the opposite, where "feathering their nest" seems to be the single-minded mission in life. Even for the lunar, space shuttle, and ISS astronauts, the destinations are calculated to the micrometer, and a guarantee of safe return home is absolutely an objective, so what is that Nat Geo writer talking about!

How many people just get into their cars, SUVs, or RVs – vehicles designed and equipped for the purpose! – and just head off into the sunset – such as might be imagined as a scaled-down equivalent to the Star Trek voyages? With no particular destination in mind and with no intention of stopping. The answer is simple: **people have homes and we like to be in them.** And not just as protection from the elements – it's more complicated – because it affords us a *guarantee* of safety, comfort, convenience, fun, pleasure, and fulfillment. And this is, obviously and evidently, what people love. A Star Trek or nomadic existence has none of that. We place tremendous value in our homes, they are the center of our existence, and their market value is the single biggest driver of our economy. I may be intrigued by the desert, but that doesn't mean I want to live there. I recently drove through the desert states New Mexico and Arizona. I didn't stop. *"There is nothing in the desert, and no man needs nothing."* Many are intrigued by the unknown, but only from a safe distance (between your favorite chair and the TV screen) because

People really HATE the unknown and uncertainty and drama – that is our nature, and NOT to be adventurous.

What do we do every day after work? We get into our cars and go home. But we don't just go home – we race home! Gotta get home, gotta get home. And someone's probably waiting for us and god forbid we're even a minute late or the questions begin and the jealousies and envies spring forth! **People have homes and we like to be in them.**

Look how we are with the weather. Weather forecasting is one of the most high-tech things we have. People go crazy if they don't know what the weather will be tomorrow! The weather! Even though we all already know approximately what the weather will be tomorrow (like today, of course) we have to know exactly what it will be – and we've invented sophisticated computer and satellite technology just for this purpose! The weather! People are afraid of even that slight uncertainty. So no one is going to forsake their happy abode to get on a spaceship literally headed for the unknown, and space tourism promoters are delusional if they think there's a market for this. Only a fool would invest in such an industry thinking there was and expect a profit. Even the airlines don't make a profit, losing millions every year – and we all love to fly! Every part of planet Earth is serviced by a colossal airline industry and it still doesn't make a dime of profit! A space tourism industry isn't just economically unfeasible, it's idiotic and misguided for any number of other reasons, not the least of which is that *we're already in space flying on a giant Goldilocks planet – aren't you having fun already!*

Teleportation (along with warp drive) is another one of those sci-fi concepts that has succeeded in intriguing, thanks to the popularity of *Star Trek*, but is one of those things that will always remain – for practical purposes – or any purposes! – sci-fi. It's one thing to teleport an electron a few feet or miles, which some quantum physicists claim to have done. Or even an apple. Just think if we could teleport a bag of groceries to a starving tribe in Africa. Or an entire grocery store! But people? Well, OK – if dreams of teleporting ourselves to a distant Earth-like planet (or bags of groceries to Africa, which would, obviously, save on shipping costs – or would it?) inspire people to become scientists – and leads to cures for cancer or world peace – that may be its ultimate purpose.

35

The Mars Adventure Paradox

Man is a hugely complicated creature with enormous biological, spiritual, and metaphysical needs and appetites that must be fulfilled, but which could never be fulfilled on Mars.

Even the most adventurous are really about 1% adventurous. 99% of the time they're like the rest of us.

GOING TO MARS WOULD BE AN ADVENTURE, BUT ONLY IN A TRULY quixotic sense. Earth is the adventure! Where the opportunities for adventure and thrills are unlimited and wide open! To think that Mars is for adventurers is part of the Mars advocate insanity! And we obviously have the means to exploit every opportunity for high adventure right here on Earth, with our cars, trains, airplanes, bikes, or just plain walking. You want adventure? Walk around the globe! Move to a different country! Switch to a different religion! Get married, get divorced, become a polygamist, make a baby, open a franchise, quit your job and live on the street, don't quit your job but live on the street anyway, hike to the southern tip of South America, do it backwards. Your options... are infinite! Buy a convertible and drive with the top down! Jump out of an airplane! Have promiscuous sex! You won't be able to do any of these things on Mars! There will be no adventure on Mars.

The Spanish monarchy didn't fund Columbus's voyage to America out of agreement that it would be "adventurous." Nor did Christopher Columbus use that argument as it was probably the last thing on his mind (as opposed to the personal riches and royal titles to be gained) – so why do the Mars advocates use that argument to justify going to Mars?

Regarding the first suggestion, to walk around the globe – imagine that! Talk about adventure! Call it the "walking or floating" tour. Plot a course to circumnavigate the globe *on foot*, where the route would correspond to any circumference of the globe, or *great circle*, such that you'd cross the equator twice before returning to your starting point, but in any direction (the equator is one example of a great circle), the idea being to maximize the length of the path you'd take around the globe – that way you could say you really did something. The only rule would be "if you're on land you're on foot," no cars, bicycles, skateboards or anything that rolls. For crossing water you'd be allowed to use any kind of non-machine powered boat or floating device, whether it's a kayak, rowboat, replica of the Santa Maria, or whatever might be available. The "walking or floating" tour. Otherwise there'd be no rules. You could wear whatever you wanted, eat whatever you wanted and stay at fancy hotels along the way, stopping to enjoy local customs and hotspots – which would, of course, be a major part of the adventure. Sound crazy? Not as crazy (by which I mean suicidal) as going to Mars! The only reason anyone would go to Mars is because they HAVE NO IMAGINATION. Certainly walking around the globe – planet Earth – would be an adventure.

On Mars, man would need to live in a completely artificial environment, since the Martian environment is completely deadly. Everything would be artificial – the pleasures, the pains – and nothing real. The scope of this artificiality may be unlimited, with no pleasure to be obtained from anything we're familiar with – food, sport, the usual pastimes, sex... The food would be an artificially flavored substrate, and artificial insemination or test tube babies may become the rule, so even the simple joy of natural reproduction would be gone. It would be an entirely fake existence! I don't see how any of this fits into a life of adventure – do you?

This does not prove that living on Mars would be impossible – but what is life without these fundamental pleasures which to a very large degree define our daily lives rich and poor, smart and stupid equally. These are the constituent things of our lives! These are the things that make us human! Unless you believe, among other things, that real sex and the sexual agenda itself should become obsolete.

Mars – the New Prerogative

If only because the odds of being successfully vetted for astronaut training are less than the odds of winning the lottery – much less! – it's ridiculous that anyone would now think of going to Mars as a *prerogative*, which is the implication made by Elon Musk, that anyone can go to Mars. Or for any youngster to think that what might have been a whimsical fantasy for previous generations...would now be a practical ambition!

SPACE TOURISM – THE NEW SNAKE OIL

The role of the astronaut is outdated, leading to fewer rockets being built, so by opening the doors to the masses – and calling them "space tourists" – one might expect some potential for a rocket building industry. After all, if you're on a rocket, it makes little difference whether you were recruited from the military – or some rich tycoon who doesn't know what to do with his money. They're both astronauts! Or they're both space tourists! What's the difference? In this light, space tourism is not a new thing, but an old thing, which means **Neil Armstrong was the first space tourist.** But why limit this distinction to astronauts, in which case anyone on an airplane might be a space tourist! But certainly "space tourist" is a contrivance and space tourism the new snake oil.

> **Why do we create the false dichotomy of "this" space and "that" space? Space is space and there's only one space and we're all in space already – in which case we're all already space tourists – on a grand tour through the galaxy in the full, literal sense, each of us in our own front row seat.**

Why does anyone need to be – of all things – on a rocket flying away from Earth to be a *space tourist*? Even if we bought into it for a second, the doors aren't really thrown wide open – except for the rich! But why would a rich person want to go into space? What is a rich person but someone who has mastered the rules of the game of being an earthling, and why would such a person want to give up that game for another? There is no such person – at least not in the numbers needed to support a rocket building industry. So it would seem that, despite the appearance of throwing the doors wide open, so-called space tourism is actually MORE selective – because there are fewer rich people who want to go into space as a tourist than people who might qualify through military channels. So if you want to go into space, your odds may be better joining the military and going that route than thinking you might have this prerogative as an ordinary working stiff.

36

Genetic Adaptation

‖‖

That man will "just evolve" is actually one of the more mind-blowing assumptions made by the Mars advocates, since it is more likely that, on Mars, man will "just die." Evolution allows for possibilities, not certainties. Evolution increases the odds that something will survive – but not necessarily us. Years after we attempt a colony on Mars, it may be home to numerous living species – but with no trace of humans whatsoever!

GIVEN THE EXTREMES FOUND WITHIN THE COSMOS, ONLY A FRACtion of which can be found within our own solar system, and how specialized our own requirements are, and that the case for adaptation is largely overstated...what might be the odds that there is another planet that could support our existence? Considering that there are billions of planetary systems within our galaxy, this might suggest the odds would be favorable to finding an Earth equivalent. It's impossible to know, or even guess – maybe 1 or 2 per thousand exoplanets at most, and one per 100,000 exoplanets at least. Not necessarily already supporting a living ecosystem, but with temperature, pressure, oxygen, water, gravity, and radiation within limits that could support human life as we know it – and preferably without the need for a spacesuit. Not having conditions that would require us to devolve into some other form of life, otherwise why not just shoot a rocket full of dogs and cats into the abyss with fingers crossed? It may make more sense to send up monkeys – and hope they evolve – than humans, knowing they'll devolve. But this seems overly logical, because it seems that living things don't devolve – surely that's just a play on words – and that evolution (or adaptation) doesn't exist as a form of crisis management – as if a living thing, taken from its element and introduced into another, would spontaneously transform into something else – rather it would simply die. (I won't treat humans as a special exception such as might be provided by wearing a spacesuit, since we're subject to the same terms of evolution as any other living thing.)

Evolution (or adaptation) only applies to living things that are there from the beginning – in its resident ecosystem – and exists as a form of synchrony. This means if we move to another planet it must have the right conditions from the start – a living thing can't just step in from the outside at some random moment and expect to be taken in, as if evolution is a kind heart that's around every corner just waiting for us, to bestow us with eternity. So, if the point of going to Mars is to preserve the integrity of mankind "as we know it" – this would be impossible on Mars from this evolutionary standpoint.

The idea that man could plan or arrange his own evolution (as a survival mechanism on Mars) might be more absurd than we think, like a subject in a painting revising itself, because evolution is not so much a master plan as it is a retrospective, with its trails leading from the past to the present. Our success as a species has not been due to us steering it (as might be imagined with genetic engineering) so how could this be the case going forward – on Earth or on Mars – but much less on Mars where everything is against us starting out.

Even genetic engineering would not likely alter the weighty business of evolution, our experiments amounting to nothing more than child's play.

The Mars advocates are undeterred by the fact that Mars is, by any earthly standard, unlivable, and over the years have come up with some pretty wild ideas how we might adapt to Mars' hostile and unlivable environment, such as engineering gill-like structures to more readily extract oxygen from the CO_2 rich atmosphere (Scientific American, January 2013), or "toughened skin to resist the huge amounts of radiation and low pressure" we'd be exposed to. Why don't we do these things now to make our bodies even more suited to living on Earth? Toughen up our skin to make it more resistant to radiation damage...grow gill-like structures so we can live underwater...grow eyes on the back of our heads...grow another set of arms – what's keeping us from creating such wonderful "adaptations" here on Earth NOW?! If we can be so clever about adapting to living on Mars, why don't we apply that same inge-nuity to accelerate our adaptation to Earth – because we certainly have a long way to go just in our capacity as Earthlings. Who hasn't wished they had another set of arms, or eyes in the back of their heads?

What is it about living on Mars that will – finally – bring out the seething, bubbling ingenuity that – the Mars advocates would have us believe – lies dormant as we continue to shuffle around on Earth?

It's amazing how these things will just naturally emerge on Mars, when we will suddenly have a mastery of gene manipulation, etc., as if Mars is the, until now only dreamed of, *trigger* that will unleash all this potential...to be what?

What will we be after all this? Supermen? Immortal Super Beings? Or just more of the same – shuffling around but on Mars instead of Earth.

And why do the Mars advocates think that Mars could possibly provide a resource-rich proving ground for the research and development that would be necessary for all this? When is someone going to call these guys out – most of their assumptions are asinine! They're stupid! They won't work! Why is it that when you couple an engineering degree with an inflated ego and a little imagination you end up with *messianic A-holes?!* Which is what the Mars advocates and engineers are – some of them.

STEADY-STATE CRISIS MODE

People on Mars will succumb to exhaustion from their perpetual state of problem solving and trouble shooting. The human brain isn't designed for that. No human functions this way. Living creatures spend most of their time in a resting state and a steady-state crisis mode will lead to a battle fatigue from which there will not only be no relief, but no HOPE for relief.

Is life worth living when you're always in survival mode? To be preoccupied with just surviving? Like being in a perpetual state of "at the top of Everest."

Can any human maintain a lifestyle, on a 24/7 hour basis, live in a world, where everything is rationed, measured, and counted, and knowing that if the supply runs out YOU WILL DIE? Where even the means to replenish some material is limited and RUNNING OUT or BROKEN? Can a human being maintain this degree of attention, of care – living in a steady-state crisis mode – indefinitely or on a permanent basis? The human brain isn't wired to do this – to literally LIVE IN FEAR. We evolved in a world where there is an excess of the things that are vital to us, where our brains don't need to maintain a perpetual degree of critical attention, perform ongoing "quantitative analysis" just to satisfy the simplest biological needs – breathing, drinking, eating, reproducing – *which are all instinctive and not calculated.* We didn't evolve to even be conscious of these things, they just happen.

This "form follows function" is reflected in not just the wiring but the structures of our brain – which evolved over hundreds of thousands of years for the singular purpose of living on Earth. From this perspective, there are no long term prospects for man's survival on Mars.

Because you can't just rewire the brain, or redesign it, and any potential for this – as our genes experiment with survival tricks – would surely be defeated in an environment as extremely incompatible with human life as Mars.

ADAPTATION HOARDS HINDSIGHT RATHER THAN FORESIGHT

It's too bad that "adaptation hoards hindsight rather than foresight," otherwise the Mars mission would be easier. When I say "adaptation" I don't mean the more vernacular "getting used to it." Besides adaptation, biologists and psychologists use the terms "flexibility," "acclimatization," and "learning" to describe the different ways we adapt to our environment – how we're able to live in an environment that changes over time, or move between different environments. It's important that we understand these distinctions, because they suggest limits that will be imposed on our ability to live on Mars, limits that can't be defeated by sheer will power or training. We have to take these limits very seriously and not ignore or deny them. When geneticists say "adaptation" they're referring to natural selections that occur through *genetic mutations*.

I would not entrust the survival of man on Mars to this scientific meaning of adaptation at all, which would require that some characteristics already exist in man that allow him to survive right from the beginning.

That is, some individuals would die off relatively soon, while some others *might* live to pass on their genes to the next generation, some members of which would die, while some others *might* live long enough to pass on their genes to the next generation, and so on, each generation becoming more adapted to Mars than the previous, with unfit individuals dying off, but this scenario would require a large population starting out, thousands, to select for survival characteristics, and have a high death rate. But since plans are to send up only a few at first, this adaptation scenario would be impossible – nor is a high death rate part of the plan.

Alternatively, if one is expecting a quick genetic adaptation – that is a dream that would just not come true because, again,

Species don't adapt quickly, not within 1 generation, nor in just 1-2 generations – they adapt slowly to small changes if they don't die first because genetic mutations that occur with each generation need sufficient time to pass onto the next generation.

The mortality rate should decrease as natural selection and adaptation continue, over generations. But even this is theoretical because evolution is a retrospective, not something that is in plain sight. Any visible adaptation would probably first be regarded as a birth *defect*, rather than as evolution working its magic.

What the Mars advocates seem to be expecting is a major and spontaneous reshuffling of man's genotype, which is impossible.

This expectation would be less ridiculous if man's genetics were *polyploid* – where mutations and therefore genetic diversity occur at a higher rate – but man's genetics are diploid, not polyploid.

That man will "just evolve" is actually one of the more mind-blowing assumptions made by the Mars advocates, since it is more likely that, on Mars, man will "just die." Evolution allows for possibilities, not certainties. Evolution increases the odds that something will survive – but not necessarily us.

Evolution favors life but this favoritism is certainly not species specific. We are here *despite* the odds. Every year, an estimated 10,000 to 100,000 species die off. Every year! Clearly adaptation is not a trustworthy ally!

Years after we attempt a colony, Mars may be home to numerous living species – but with no trace of humans whatsoever!

So, adaptation can't be relied upon to guarantee man's survival on Mars – it's not a sure thing – as evidenced by countless species (on Earth) that have become extinct with even slight changes to their environment. Adaptation – in consideration of the fact that it hoards hindsight – can't be planned or strategized. The results of adaptation – survival of a species – are an accident even when the species is lucky enough to have time on its side – which will *not* be the case on Mars.

For those who think evolution is a tool that we can control, it's really the opposite: we are the result, one of countless possible variations – that evolution has devised.

For us to control evolution is like someone in a painting thinking they can control the brush strokes. The field of genetic engineering attempts to tinker with this grand design, so that we might control evolution, but this would not change the general nature of evolution, nor have any bearing on my statements about adaptation. Bioengineers will *not* likely control the future of humans as a species, not because it may be unethical, but because of the sheer momentum of the human gene pool.

We already control the genetic future of humans, even as a species, and completely without genetic engineering – by healthy couples simply having many children and unhealthy couples having few or no children. But it seems evolution already favors this strategy and won't be second guessed!

The adaptation process doesn't know what we want, its effects aren't intended to be beautiful or elegant – it's the genes throwing their own little dice – and usually are a compromise between the species – such as future Martians – and nature. *"Adaptation hoards hindsight rather than foresight."* So don't be surprised at what might happen as a result of our beloved adaptation: Martians might sprout gills, and not in a place we expect or want – Horrors! But one thing is for sure: people on Mars won't adapt wearing spacesuits or in Mars-proof buildings – this would leave them attached to life support devices forever.

Yes, sheltering man from the Martian environment would actually hinder naturally selected adaptations that might otherwise occur, while genetic engineering would rely on procedures best developed on Mars but which couldn't exist without much "trial and error" and which of course don't exist. An alternative to this would be for humans to be "genetically engineered" on Earth, suited to living on Mars – and in the meantime somehow kept alive on Earth. In another scenario, in a colony of 80,000, and which I'll presume wouldn't be overly protected while on Mars, so that adaptations could occur more naturally, you'd have a large amount of genetic variability, to maximize the emergence of adaptations and therefore viable individuals – but that's some *crazy experiment*, with a potentially high death rate – we might be deterred by *ethical issues*. Species adaptation to a new planet that has many hostile influences would be very slow, span many generations, even centuries, since the adaptations are genetic and need time to pass onto successive generations, during which time many – if not all – could die if sufficient adaptations don't emerge. Alternatively, with biological engineering techniques – surgery – you might circumvent the high death rate incurred with Mars based trans-generational adaptations (i.e., natural selection) but this would lack the advantage of being heritable and would have to be performed on each and every resident of Mars.

The most obvious problem with relying on natural selection and adaptation to survive on Mars is that we would most likely first be defeated by the very fatal limiting factors concerning primarily oxygen, water, and food before the generations long natural selection process would have enough time to succeed. Even if we could accelerate the maturation process of women, such that they could bear young at the age of ten – but which might accelerate the aging process such that we could end up short-circuiting the human life span by many years, and which might cause an outrage among earthlings – and go so far as to exploit woman for their breeding capabilities, this would offer probably only a slight advantage over the normal life cycle. Another option would be surrogacy, with adult Martian women recruited en masse to carry fetuses that arise through test-tube fertilization, with eggs being selectively harvested and processed from females as soon as they mature. This would minimize the length of a generation, which could be further reduced if ways were developed to either accelerate the maturation of human eggs, or to somehow modify immature eggs so that they could be fertilized – any way to accelerate or circumvent the normal life cycle and therefore speed up adaptation. Cloning isn't an option because that's asexual and therefore doesn't follow the principles of natural selection as does sexual reproduction, which is the very thing we would be trying to exploit in our mission to achieve some viable blend of adaptations – in perhaps the greatest experiment of all time – as we attempt to "save the human race."

As previously mentioned, flexibility and acclimatization will further determine how well anyone adapts to living on Mars, but these are not genetic, not inherited, and dependent on each individual, so an individual that does just fine in this regard...may have kids that just don't.

These are the pretty details the Mars advocates don't describe on their websites or books – what the "science writers" blithely disregard.

IF THE WONDER OF "ADAPTATION" IS SO TRUSTWORTHY THEN THIS magic should lend itself to issues here on Earth, like dietary fat metabolism, among other modern "defects." Adaptations take generations to work, being genetic, but...we shouldn't need to worry, now that we've been reminded of the phenomenon of adaptation, about our saturated fat intake and which is at least partly to blame for the disturbingly high incidence of obesity. We'll adapt! Meanwhile, people will die, but isn't this a worthy price to pay – the inevitable adaptation will allow us to down Big Macs like they were rice cakes. This example clearly demonstrates how slow and painful the process – this adaptation – can be.

> **But we have significance as individuals, moreso (the way we're playing it now) than as a species. Should this balance be preserved on Mars? – in which case it seems adaptation would be stymied. Or should the individual be subjugated – even sacrificed – for the advancement of the species, where the focus shifts from survival of the individual to that of the species?**

Believers in adaptation – the Mars advocates – should say yes. And we should all apply this same free thinking innovative spirit to life on Earth![1]

This is one circumstance of Mars that the Mars advocates could be exploiting: the complete absence of other creatures on Mars – that would only serve to compete with us for resources and space – might just work more to our advantage than to our disadvantage, and completely nullify the threat of interplanetary crossbreeding or messing up what elaborate ecosystems might otherwise exist. We certainly don't want to compete with some dominant life form that turns out to be territorial. Or to disrupt its evolution. Whatever life forms that may have existed, in some imagined watery world of the past, were obviously too fragile to survive

----

1   It would seem that obesity, from the standpoint of species survival, isn't really a public health issue at all, in light of the fact that it doesn't seem to affect the longevity of our species. As long as people live long enough to reproduce and then raise their children, it may actually be better that people die as soon as possible after this, otherwise they are simply consuming resources that might be better preserved for future generations (and creating a lot of waste in the meantime). And on a planet where all the elements we need to live exist in huge excess, including food (meaning anything that is edible), and delivery of this food to our stomachs has become so efficient, it stands to reason that people would tend to be overweight. Granted 20,000 die every day of starvation, this seems to be in very isolated pockets, but the fact that our world population strongly leans toward overpopulation suggests that even this – malnutrition and starvation – may not be the problem we make it out to be.

beyond that past – *which further demonstrates that natural selection is no guarantee* – assuming this theory could even be extended beyond Earth. So the fact that Mars isn't already crawling with life might be a significant advantage – if it weren't for all the other overbearing facts crushing the Mars advocates.

> **On the other hand, humans evolved among thousands of other species of microbes, plants and animals. Will the lack of this diversity of species on Mars be a problem for our future evolution on Mars? By the same token, if we attempt to transfer plant and animal species from Earth, will this backfire in some weird zoological nightmare, with species mutating Dr. Moreau style, the behavior of DNA itself gone berserk?**

Speaking of maladaptations, the "science writers" claim that it's the nature of man to explore. Not really, if "natural" refers to a tendency of the majority. In fact, an intense need to explore (as opposed to simple curiosity) seems deviant, oddball, even sociopathic. Many if not most extreme explorers/adventurers are genetic dead ends, someone's uncle since most wind up dead at a young age. A trait is a maladaptation if it interferes with the ability to pass on one's DNA, which happens a lot to the extreme explorers/adventurers, and they wind up the celebrated uncle. So the gene for exploring might be a bona fide maladaptation. We sometimes envy their "exciting lives" but you've passed on your DNA and they have not – what does that tell you? In the same light, through the brute force of logic, not only is the desire that we should go to Mars, and the belief that we can survive there, misguided...it's a *maladaptation.*

No matter how bad things get on Earth (disease, asteroids, epic floods, depletion of ozone, nuclear fallout) ...won't we just adapt?

> **It seems that the Mars advocates want to bring on the Hell that they're so afraid of – call it the Mars Advocate Paradox, and which seems to be the crux of their misguided logic and folly. Why do the Mars advocates think that natural selection has stopped working on Earth – but will work fabulously on Mars?**

I'll reduce it even further: Isn't it better to play the *long* odds of adapting to a *possible* Hell on Earth (which is also much cheaper and will provide for a large base population from which to reconstitute civilization, if needed) than the *even longer* odds of adapting to the *certain* Hell on Mars (which is much more costly, would provide an insufficiently diverse base population, and in the meantime allow only for a miserable existence)? It seems like a no-brainer: stay on Earth and enjoy the Garden Party.

37

Interplanetary Germ Traffic

This will pose an unending nightmare to Martian society, as Earth immigrants continue to arrive and take up residence in the shared, communal, and very much enclosed living areas.

The diseases that each world will have will impose serious limits on whatever commerce or "space tourism" might be imagined between the two.

Any future dietary supplement regimen for people on Mars would need to include not only vitamins, but immunosuppressants.

IT MAY TURN OUT THAT "LANDING A HEAVY SPACECRAFT GENTLY" or "the return trip" – or lack of oxygen and water – may be nothing compared to the challenges posed by...the Human Immune System. Talk about overly complicated systems! It is arguably too complicated for its own good – and sometimes seems to violate the principles of so-called intelligent design, as with the many autoimmune diseases (in which a person becomes "programmed" to self-destruct) – and is increasingly implicated in diseases and health conditions that appear to be more straightforward. Even separate from any Martian influences, the human immune system is a huge wild card, and figures in many apocalyptic and extinction scenarios predicted for Earth and would therefore present an equivalent threat to a Martian colony *in and of itself even without the ongoing influx or influence of Earth germs.* Let's put this – *Interplanetary germ traffic* – in our new vernacular. This may present an even bigger obstacle to interplanetary commerce or other interactions between Earth and Mars than the more obvious ones that preoccupy the rocket scientists – and other "specialists" that tend *not* to include immunologists – some that *might* be "engineered away" if only slightly – making a trip to Mars more one-way than anyone would like to imagine and forcing a self-sufficiency on any Martian colony...that would be truly impressive.

As we evolved, our physiologies became intertwined with the universe of microorganisms, which affects even our moods, and scientists have only recently discovered just how profoundly complicated this relationship is. Ten percent of our body weight is microorganisms, including beneficial bacteria, which are crucial to our existence – at least on Earth. The effects on this dynamic of the sterile Martian environment will be a big question, with man's biology and psychology being stunted in ways we can't imagine as man attempts to become a permanent resident on Mars.

In science-fiction (always a good reference point), alien species that visit Earth are often done in by some unsuspected adversity, such as microorganisms. Will this same logic be at work with Humans on Mars? It's hard to believe it wouldn't be. Who knows how the immune systems of those moving to Mars will react to all the various Martian elements, the red dust in particular, which even in minute quantities may build up in the lungs and be the Martian equivalent of mesothelioma (the asbestos related lung disorder) or pneumoconiosis (black lung disease, which affects primarily coal miners), both of which are deadly and incurable. Apollo 17 astronaut Harrison Schmitt had a severe allergic reaction to moon dust during his lunar mission in 1972 [http://www.wired.com/science/space/news/2005/04/67110]. "Dust is the No. 1 environmental problem on the Moon," reported Schmitt in an interview. "We need to understand what the (biological) effects are, because there's always the possibility that engineering might fail." Lunar dust is unlike Earth dust, littered with bonded shards of glass and minerals known as ag-glutinates, a phenomenon that doesn't occur on Earth. It's like Velcro, having sharp angles, with arms that stick out and little hooks, according to David McKay, as chief scientist for astrobiology at NASA's Johnson Space Center – making it easy for the particles to get deep into the lungs and stay there. Scientists believe that the human body may not be able to expel this "dust" if inhaled, and lead to fatal lung diseases similar to silicosis.

We might as well expect similar problems with Mars' red dust – which literally covers the entire planet and fills the air during its infa-mous planet-wide dust storms that can leave dust suspended in the thin atmosphere for years, with the possible formation of hydrogen peroxide and other corrosive chemicals that fall to the Martian surface as a sort of toxic snow. Add to this perchlorate, a chemical implicated in certain soil samples taken by the Mars rover Curiosity in a study led by Laurie Leshin, dean of science at Rensselaer Polytechnic Institute in New York. Perchlorates are toxic to humans, and known to cause thyroid problems. Sounds wonderful. The Martian red dust may be the No.1 environmen-tal problem on Mars, with no end to the problems it may cause.

Reality: research indicates that living in space weakens astronauts' immune systems. According to the European Space Agency (reported March 26, 2013 by ESA at their website), scientists from the University of Sassari, ETH Zürich's Space Biology Group and Zero g-LifeTec have been studying monocytes, a type of white blood cell in the immune system, in microgravity aboard the ISS, and discovered the monocytes have altered cytoskeletons, which reduced their motility. That change, according to the ESA researchers, may partly explain astronauts' weak-ened immune systems. Just another problem we can engineer away?

The diseases that each world will have will impose serious limits on whatever commerce or "space tourism" might be imagined between the two. Immune systems of those born on Mars and later generations will be much less developed than the first generation of Earth immigrants, since Mars is, by all measures, sterile, so as soon as the first Martian is born if not before, one of the first things residents of Mars will have to do is set up zones of quarantine for new arrivals from Earth, since they'll be carrying germs for which those born on Mars have no resistance. This will pose an unending nightmare to Martian society as Earth immigrants continue to arrive and take up residence in the shared, communal, and very much enclosed living areas.

> **From the perspective of anyone on Earth, it would be best to consider any Martian colony to be in a permanent state of quarantine.**

Then there's the separate issue of Martians visiting Earth, and perhaps back again, travelling from one biosphere to the other. It may be that those born on Mars may be so unlike ourselves, in terms of the germs in and on their bodies, that visiting or moving to Earth would be impossible without posing serious risk to either themselves or us. Some believe that an alien invasion of Earth would be defeated by our microbes, our diseases – but it seems the problem would be axiomatic for future Martians visiting or moving to Earth, who may be only slightly better prepared (than an actual alien), and need to reside in a state of quarantine while on Earth, where something as common as a cold virus could be fatal to a visiting Martian and have a completely different pathology than for us. Future Martians may be forbidden from visiting Earth for this reason alone – what a pity! Oh, to be a Martian, what a doomed fate...!

38

Extinction

The Mars advocate ideology may itself bring about a doomsday scenario that didn't already exist, or other futures that are completely contrary to the path taken by evolution.

If the evolution and rise of intelligent creatures is one of the objectives of the universe – throughout the universe and not just on Earth – and we assume this to be axiomatic...then it wouldn't matter if any one of them, or several, took a wrong turn somewhere and became extinct. The natural IQ of the universe would not be diminished by our extinction or the end of the world – the universe would not need to "start over again." All of which means we don't need to worry, for the sake of a self-conscious universe, if we become extinct. This completely nullifies any sense of criticality about getting to Mars!

ANYWAY, IF ONE OF THE REASONS FOR MOVING TO MARS IS TO save ourselves from an extinction level event on Earth, this begs the question: will being on Mars necessarily save us from extinction level events – even ones that occur on Earth? Many extinction level events that could occur on Earth would also affect Mars, such as Earth being struck by a very large meteor.

It is also true that many of the apocalyptic scenarios that imperil Earth, if unlikely now or the foreseeable future, could also occur on Mars, such as being hit by a comet or asteroid, indeed may be more likely on Mars, since Mars is closer to the co-called asteroid belt that exists between it and Jupiter. Mars is also closer to the Kuiper Belt (an area extending from just beyond Neptune) and the further out Oort Cloud (a more theoretical area thought to extend to the edge of our solar system), from which all asteroids, comets, meteors, and similar flying things originate, besides the asteroid belt. Mars, to a degree, actually shields the Earth from asteroid hits. This diminishes the Mars advocate "apocalypse by asteroid" argument, which claims Mars is safer than Earth regarding this particular issue so in this regard we are safer *on Earth*.

The Yellowstone supervolcano is high on the list of Mars advocate fears, but the same people who would be killed in such an event...would be killed even if there was a colony on Mars. Contrary to some Mars advocate fears, the Yellowstone supervolcano does not pose a risk of extinction to humans. "Inhospitable conditions would persist in the Midwestern U.S. for about a decade, with new vegetation starting in about that time," says Stephen Self, director of the Volcano Dynamics Group at the Open University, United Kingdom (mnn.com, June 2012). This would therefore not lead to the extinction of humans, quite to the contrary, and advance warnings in the immediate vicinity would allow many to avoid being killed in the eruption itself. So this too does not make a strong case for Mars. Self continues: "recent research shows the global impacts of supervolcanoes are less severe than scientists once thought, and a Yellowstone supereruption might be especially *unimposing* because its magma contains minimal sulfur." "As for the rest of the world, it would face a few years of mild climate change caused by the supereruption's ash cloud, which would wrap around the globe, but casting Earth in shadow for only a few *days*." "Scientists don't see any evidence in the geologic record of mass extinctions coinciding with supereruptions, and based on new models, they don't predict extinctions to result from such geologic events in the future." "The last time Yellowstone erupted, no extinctions took place," said Michael Rampino, a biologist and geologist at New York University. "Supereruptions are not extinction level events."

But on Mars, conditions being what they are, we would, relatively speaking, be on the verge of a species-level extinction event from the outset, whereas the odds of a bona fide extinction level event occurring on Earth are pretty slim, since there are so many of us, we're widely distributed, and we have sophisticated means to predict and avoid danger. Let's consider a scenario where 99.99 percent of the human population was wiped out by some catastrophe. That would still leave 800,000 humans on Earth – which would probably be self-sufficient in one form or another, and our species would survive. Several entire cities would probably be largely unaffected. But if there were 80,000 in a Mars colony, and 99.99 percent were wiped out, that would leave only 8 Martians shivering in the cold – not enough to start out again. And something that would be barely insulting to Earth or an earthling might be all it takes to wipe out 99.99 percent of a Mars colony. Put in this bright light, a Martian colony is looking less and less worthwhile – more like a dead-end. For a Mars colony to be considered a bona fide hedge against our extinction, it might need to be large enough to survive a catastrophe of this order of magnitude – such that the 0.01 percent that survives is at least, say, 400. Such predictions are completely speculative, of course, but that would require a base population of 4 million people on Mars – to provide a colony of a size sufficient not just to be self-sustaining but to survive a near species-level extinction event – which is, of course, the whole point, right? So, regarding a colony of 80,000 – ridiculous enough, but 4 million? This seems even more impossible, even in the long term with rockets 10 times faster than the present – "safe and ordinary" technology of this nature seems highly unlikely but this is, of course, what it would take. And you couldn't simply ramp up to 4 million over decades, you would need a starting point of 4 million – which is of course impossible when talking about a human population on Mars – it's not like we're bacteria in a Petri dish. Better to forget Mars and rapidly prototype the next gen "almost the speed of light" rockets to get to a more Earth-like planet in one of the neighboring solar systems, where a near species-level extinction event wouldn't be as likely (since it would be intrinsically more habitable than Mars to begin with) and we would only need to send a few hundred – exponential increases in distance put extreme limits on the numbers that can be transported safely even with exponentially faster rockets, since costs will also probably rise exponentially and where "safe and ordinary" travel may be too much to ask.

But would any other Earth-like planet not already be seething with life: from microbes to intelligent civilizations – perhaps like us on a curve toward self-destruction – or paranoid of alien visitors come to steal their resources or dissect their private parts?

But no – certainly any Goldilocks planet we find would be a Garden of Eden ripe for the picking! Even better: refocus our priorities toward Earth, realize that colonizing outer space is out of our reach, realize that Earth is the Goldilocks planet we're searching for – we're on it! – that any Earth equivalent would probably have the same or similar adversities as Earth and that whether we're on Earth or some Earth equivalent that it – inevitably – comes down to humans being the smart species we claim to be. Ultimately, we're already on the planet most favorable to our species survival – which is, of course, axiomatic.

So it seems a colony on Mars doesn't really increase the odds of our long term survival if an Earth Armageddon is what we're avoiding. If Earth's waters become polluted, fish aren't better off on land – the manner and location of death are simply shifted, but where all doors lead to the same thing, what's the difference? Likewise, a colony of people on Mars doesn't really increase the odds of our survival – the manner and location of death are simply shifted.

The simple desire for sex – a horny Martian – may put in jeopardy the colony if it's not satisfied – and this may be all it takes. A horny Martian could absolutely destroy the dynamic of any closely knit group and lead to self-destruction of the colony. Even healthy relationships eventually go bad, but think how quickly things would disintegrate if one of the colonists rapes another – where of course the rules would be completely different. Is it possible to imagine that this wouldn't happen, sooner or later?

There may be an eventual colony on Mars of about 20 people or so, but within two years or less we'll pull out fast, the thrill having gone many months earlier. Then we'll promise ourselves to reconsider in another 30-40 years, at which time the next generation will think it was their idea and start anew, and so it will continue, the never-ending cycle, but always just out of reach, while the science-fiction fantasists fuel the fire (and the dial-a-flavor will be to the Mars mission what Tang was to the Moon mission).

39

Junkyard Mars

Every landing of a spaceship will be a potential species-level extinction event. It seems unavoidable: once man steps foot onto Mars, the potential for disaster, the drama, the potential for heartbreak – the potential for horror, horrors we can't imagine – will never end.

CAN A MAN TRAVEL TO MARS? OF COURSE! CAN HE LIVE THERE, seek fulfillment, and give rise to a new generation of mankind? In other words, not just survive in some extremist, bare bones, minimalist sense...but *thrive*? Absolutely not! There will never – *can never* – be a mass migration of humankind to Mars. Not even a gradual mass migration over many years – if only because we would run out of rocket fuel first! – but this limiting factor is almost irrelevant compared to the others. If the dream of a colony is somehow preordained by the Mars advocates, even in a best case scenario Mars would become littered with dozens or even hundreds of the vehicles used to get there, the landscape dotted with junkyards of used rockets, including those that have crash landed. That's a best case scenario! A Mars advocate would say to this, OK, all the tons of metal components would be a primary source of raw materials, be reused, melted down and remanufactured – but getting to that point is not one leap but many leaps, improbability taken to the nth degree – which equates to impossible. After just a few launchings from Earth, the destination will be Junkyard Mars – but you won't see this in the gorgeous and dreamy Arizona-esque renderings provided by NASA or their SpaceX and Mars One counterparts. Such an image wouldn't be a deal breaker for the Mars advocates, but it should put things in a brighter light for them, along with futurists and dreamers among us who expect Mars to be a Nirvana or Shangri-La.

The thrill of conquest would be gone almost immediately as soon as one steps foot onto Mars, the grim reality setting in within a few days or perhaps even less, the artist-inspired expectations replaced by the sub-freezing and barren landscape. One would have to be completely devoid of their humanity to not be eventually overwhelmed by sadness and loss. The scathing reality. The stark raving mad reality. And you couldn't just turn around and go back.

We would be creating a lot of waste, of course – like on Earth. There are prohibitions against littering in space, for those on the ISS, since anything they eject would simply go into orbit and pose a collision risk for everything else in orbit (at that or similar altitude), but once we step foot onto Mars, how long would it be before we start tossing things over our shoulder? Probably right away, immediately. There would be absolutely no delay from the day we set down our garbage and trash producing selves to the first piles of trash that would only grow and grow – Junkyard Mars – and defined primarily by the junked and crashed rockets that would accumulate over time. And what a sight for each new arrival, the colony looking more like a rocket graveyard than anything else – welcome to Mars! From the utopia of Earth to the dystopia of Mars – are you in shock? Every rocket from Earth would be headed for the same general target location, since there would only be the one colony. Otherwise, additional colonies wouldn't begin too far from the first – walking distance – since the colonists would be enormously dependent upon each other – the greater good would be entirely dependent upon a unified whole. There would, of course, be no infrastructure for waste removal or recycling, and Mars is just desert so what's the difference, it's not as if anyone would be worried about curb appeal. Man is profligate – the inevitability that this would also be true on Mars – and beyond – is unequivocal.

Our presence on Mars would, of course, obfuscate the results of ongoing science projects, at least those intended to discover Martian life or compounds and molecules we associate with life, since our mere presence would be an unavoidable source of contamination. Mars would no longer be the ideal, pristine subject for scientific investigation that has endured up to this point, once we arrive – which is why it's important to learn as much as possible with the rovers, before we get there.

> **The ongoing question of whether or not life ever existed on Mars would become moot since there would be no effort to spare from the herculean, if not impossible, task of making the place "livable."**

LAUNCH WINDOW

The Earth, Mars, and all the planets are flying through space, in enormous circles around the sun, never stopping around and around, and around and around. And man flying from one to the other, then back again! Does anyone think this would be easy? Since the planets are moving at different speeds around the sun, getting from one to another requires the consideration of several things. First, it's important to minimize travel time because sitting in a rocket is both dangerous and boring – you don't want to be sitting in a rocket for 20 years if the trip can be done in a few months, because how long do you expect to live anyway? Second, fuel is heavy and what determines, to a large degree, the size of the rocket. Because of all these things, astronauts need to take the shortest route possible between here and Mars, and this means we can't launch rockets to Mars whenever we'd like to.[1] Instead, rocket launchings would be confined to what's called a launch window, which is the period of time during which a rocket can launch and expect to reach its destination quickly using the least amount of fuel. We can calculate the best possible time and place for this – the launch window – and the particular time and place in space it would "hit" Mars. As it turns out, the best time and place is only a critical few days every 26 months that would allow a rocket to reach Mars in about 8 months (and anything much faster is science-fiction). Outside this launch window it would take considerably longer to get to Mars, which would require more fuel

....

1   The distance from Earth to Mars is an average of about 140 million miles, the actual distance constantly changing according to their respective orbits around the sun.

and a bigger rocket but it's much easier if all the rockets were as small as possible and all the same size. This means once an actual rocket is on the launch platform it must launch within the scheduled launch window or not at all so if the rocket can't be launched for any reason during the launch window we must wait another 26 months. This is, of course, a great inconvenience.

> **Although the great distance between Earth and Mars guarantees the two will not crash into each other, this also makes travel to Mars inconvenient – and a terribly unlikely tourist hotspot.**

So these launch windows would be much anticipated, with great amounts of activity immediately prior to and during these precious moments. There would be much at stake – a failed mission would cost lots of money, affect the viability of the program, and put people already living on Mars and waiting for supplies at tremendous risk, so missing a launch window would have enormous ramifications. Any colony that develops on Mars would need to have enough supplies and provisions to outlast not once but twice the time between launch windows, which would of course presume a great deal of self-sufficiency right from the beginning. (Rocket launchings may be of particular interest to terrorists, and security of the launch sites and rockets would need to be invincible.) Because of the rarity of these launch windows, it stands to reason that there would be several rockets being launched at once – who knows how many! From multiple locations worldwide! Including several just for redundancy, since who knows what adverse weather or "natural disasters" may be occurring at the time to delay or prevent a launch?

Let's say during one of these launch windows that five rockets lifted off, from various countries, headed for the same destination – and which would be converging during the entire trip. This would create a risk of its own, with rockets travelling practically side by side, or in sight of one another, and which would create an extra measure of unwanted anxiety and drama. After all, you don't want one experiencing a mishap – such as exploding – with the debris crashing into another. And they'd all be headed for the same target – not just the same planet, but the same exact landing site since there would be only the one colony. This is where things would get real tricky and hardest to control. The precision of each successive rocket landing on Mars would become increasingly critical since there would be more and more junk accumulating on Mars that would need to be avoided – and you don't want a rocket crash landing on, or near, the growing colony, while at the same time you don't want astronauts to be stranded if they land too far away. For astronauts returning to Earth it doesn't really matter exactly where they land, whether on land or sea – we can just go get them but such sophisticated rescues would not be possible on Mars, where even an established colony would have far less capability to rescue crashed or stranded new arrivals. All the rockets we send to Mars need to land in the same exact area, at the location of the colony. But when you have five rockets landing within 1-5 days of each other, at the same spot, that's a lot happening at once even if all goes well!

Aircraft would need to land on a pin – this precludes the use of enormous parachutes, which would carry the craft miles off course under the sometimes windy conditions known to occur during the frequent, planet-wide dust storms.

Imagine if just one thing goes wrong, and the ensuing domino effect – every landing of a spaceship, especially the heavier ones, would be a potential species-level extinction event. It seems unavoidable: once man steps foot onto Mars, the potential for disaster, the drama, the potential for heartbreak – the potential for horror, horrors we can't imagine – will never end.

Of course you're wondering: even if a colony is self-sufficient (can 20 people on Mars be self-sufficient? 100?) can it grow at a healthy pace with an influx of people and supplies only once every 26 months? Would this schedule even allow us to maintain a stable population of the colony, since fatalities are sure to occur. We might be forced to discard the idea of an ideal launch window to some degree and just build bigger rockets, which would take more than a year of travel time. It would cost more but once a colony begins to take shape on Mars, the notion of the perfect trajectory may become quaint. Bigger rockets would serve to expand the launch window, which in turn would serve the benefit of spacing out rocket launches so they're not all landing at once on Mars, and also to launch more rockets.

AN EVEN BIGGER QUESTION THAT REMAINS IS, HOW ON EARTH will the colonists get back to Earth? I mean how on *Mars* will the colonists get back to Earth? – I forgot for a second that we're using a whole new vernacular. Granted the gravity on Mars is a third that of Earth, this is simply one variable among thousands, and I would expect the logistics to be the same as getting from Earth to Mars, involving monumental launch pads as tall as skyscrapers and 30-story and taller rocket ships. The Saturn V was 363 feet tall, as tall as a 36-story building. If this was needed to get from Earth to the Moon, I would expect something bigger would be needed to get either from Earth to Mars or Mars to Earth. But of course the Mars advocates are thinking that because of Mars' reduced gravity there might be cute little "pods" that somehow require "no effort" to lift off Mars' surface! – which is ridiculous. Furthermore, given the times and distances – and interplanetary trade that some seem to anticipate – certain efficiencies would be necessary, which of course means that ships travelling from Mars to Earth would be colossal. All the correspondingly colossal infrastructure wouldn't just magically appear on Mars – as if a Martian space industry consisting of tens of thousands of people would just pop into existence – nor do the frozen, barren landscapes of Mars exactly lend themselves to the building of such an infrastructure, especially if you only have four or five guys struggling to *breathe and eat* and can't keep the incessant dust off their visors! And the launch window for rockets launching from Mars would be equally restricted to the once every 26 months for rockets going from Earth to Mars. A lunar probe launched by the European Space Agency (ESA) in 2003 used an ion engine, which might be considered an advance, being much more fuel efficient than conventional rocket fuels, but is, alas, MUCH SLOWER – the ion-powered probe took 58 weeks to get to the Moon (where it crashed, as intended), compared to 3 days for the Saturn V rocket to make the same trip in 1969 [http://www.universetoday. com/13562]. But obviously, for manned flight, speed would be of the essence, where consideration of such things as the food supply and other things relevant to human tolerances would take a priority. Even for unmanned exploration, ion-powered rockets may be too slow to satisfy our impatient, inquiring minds.

40

Fire

From a philosophical standpoint, Mars lacks three of the four basic elements upon which everything is based: Earth, Air, Water, and Fire.

FIRE. BEAUTIFUL, WONDERFUL FIRE. IT KEEPS US WARM. FROM a philosophical standpoint, Mars lacks three of the four basic elements upon which everything is based, which are earth, air, water, and fire. Certainly from this viewpoint it is outrageous to claim that Mars is like Earth! But we can also consider these from a scientific perspective, and have already done so for air and water.

One of the other things we take for granted here on Earth, besides the air we breathe and oceans of water, is fire. Besides the wheel, the discovery of fire has been one of mankind's most important breakthroughs, and marked a turning point in our evolution, especially regarding our ability to cook food – scientists have only recently rationalized that it's easier to digest cooked food, which in turn allowed our big brains to evolve. The controlled use of fire has played an increasingly larger role in our developed society. The ability to make fire is considered to be a basic survival skill – we can't live without it. But there can be no fire on Mars, due to the almost complete lack of oxygen in the atmosphere (0.13% compared to 21% on Earth). So naturally of course this would seem to be an enormous problem!

Because of the lack of oxygen on Mars, we would have to somehow generate it. There would be a certain challenge in creating and maintaining breathable air, with respect to the ratio of 78% nitrogen, 21% oxygen, 1% argon, and traces of other gases including carbon dioxide found in Earth's air. Man can breathe other combinations, so this exact ratio isn't critical or vital, as long as the percentage of oxygen is at least around 18%.

All of which gives new meaning to "No Smoking." Any colony on Mars would be an absolutely tobacco free environment. The absence of this vice alone may make man unrecognizable! And a colonist couldn't simply go outside to smoke because the Martian air doesn't have enough oxygen to feed even a single burning cigarette. It's because of this challenge, in maintaining a habitat with breathable air, that the smoking of cigarettes would be a particular problem, which would also of course quickly upset the air quality generally, and where you couldn't just open a window for ventilation, because the dwellings would be pressurized – that would be like opening a window on a jetliner. In fact, any activity that requires ventilation would be forbidden – this implicates a surprisingly long list of things we take for granted on Earth.

The inability to make fire may be an obstacle to even a tiny colony on Mars – right from the beginning, more than we might think – where the ability to make heat will be an extreme priority on the constantly frozen Mars, besides making oxygen and finding ice to melt, and this will certainly be the case in the longer term.

> **At some point, for a colony to be successful in the way the Mars advocates imagine, Mars would need to experience an industrial revolution – of sorts – but this may be impossible without the ability to generate the high temperature fires commonly found in certain industries on Earth, such as with the mining, metals, and glass industries – or something as seemingly mundane as heat-treating stone for tools.**

Realize that cisterns of water would need to be in enclosed, heated areas to stay liquid because the surface and subsurface of Mars is always around the freezing point of water or much colder. The inability to keep large amounts of water in a readily useable liquid state would be one of the biggest obstacles to a colony of any size. Just think about that.

Liquid water. So precious – something that *does not exist* on Mars!

Even for a bare bones initial colony of just a few humans, they will want to set up a lab – but what is a lab without Bunsen burners? – which of course need oxygen. As I describe in the Microgravity chapter, everything would need to be reinvented for Mars – but it seems this would be impossible!

The Mars advocates actually expect to have mining, metals, and glass on Mars – as outlined in Robert Zubrin's crazy *The Case for Mars*. But besides lacking oxygen, Mars lacks any supply whatsoever of the fuels customarily used in such industries, such as natural gas, oil, coal, and wood. Regarding natural gas, which is primarily methane, the rovers have detected a very trace amount of methane, at the most 1.3 ppb – which is considered to be "no methane," and which is *less* than originally thought [*Science*, 9/19/13].

Because there is so little oxygen on Mars, there could be no internal combustion engines, which means that any machinery or forms of transportation would need to be entirely battery operated, and most likely recharged by the sun. But even the success of this strategy would be hindered by dust storms that occur and that can envelop the entire planet in dust for months at a time.

Fire would otherwise be implicated by the primitive lifestyle the Martians may be forced to adopt – so it would be a special challenge even to live in some primitive state on Mars.

41

No Air, No Aerodynamics

It's impossible to land cargo ships gently on Mars.

THIS FACT, BY ITSELF, REDUCES THE POTENTIAL FOR A COLONY – of any size whatsoever – to practically zero.

Large spacecraft, carrying people and supplies, would need to land gently on the hard Martian ground, but designs based on aerodynamic principles that work on Earth won't work on Mars and rocket scientists haven't figured out alternatives. Friction caused by the air slows down spacecraft returning to Earth, but this doesn't happen with the much thinner Martian air. *No air, no aerodynamics.* This is why engineers can't figure out how to land heavy spacecraft on Mars: all the science taught in schools is the science of Earth's full gravity and air! Our brains – and bodies – are *wired* for Earth's full gravity and air! And we can't SEE these things so it's hard to appreciate – especially the Mars advocates. *No air, no aerodynamics.* We don't just rely on air to breathe, but on *aerodynamics* to move through, especially when flying and to get to Mars we'd have to fly. The aerodynamics of the Martian atmosphere alone may make it impossible for creatures as big and heavy as we are (being mostly water and water is heavy) – who make big and heavy things and who want to take these big and heavy things with us to Mars – to get to Mars. If only we were like leaves blowing in the wind, we could float gently down to the surface.

This means there won't be airplanes and helicopters on Mars either, all of which represent major challenges vexing engineers and the prospects for a sophisticated colony. To truly exploit the potential of Mars our settlers would need to navigate vast areas of Mars and this would obviously be facilitated by some kind of air transportation. So starting out they might instead be using bicycles or tricycles, Segways, and "Margo sticks" – pogo sticks for Mars, pressurized to go a mile high! – and this starting out phase could last a long time.

The first three rovers bounced onto the Martian surface, eventually coming to rest, but this strategy obviously couldn't be used with people. The fourth rover was able to land gently using an elaborate, multi-stage strategy that employed an enormous parachute and tiny rockets but which only worked because the rover was relatively lightweight – but like the bouncing strategy this couldn't be used with heavier loads as would occur with people and all the supplies and materials needed to build a *new world*.

42

Primitive Beginnings

This might be one real possibility, in the form of a compromise: imagine a low-tech, machine-less existence – as before the Industrial Revolution. The Stone Ages.

WHEN YOU HEAR THAT SOMEONE IS ON LIFE SUPPORT IT USUALLY means they are at death's door, the only question being when do we pull the plug, the main concern being the potentially huge cost of keeping them alive. Yet that would be the status quo on Mars: "life support." Is this the romantic vision of Mars as portrayed in science-fiction? It hardly points to a bright future on Mars, another in the list of stark contradictions to the "we can do it" mentality of the Mars nudniks. But wait, it's worse. This "life support" status would be in effect FROM THE MOMENT OF LIFT OFF! At which point the travelers will basically be entering into an irreversible "diseased" state that only gets worse and worse as their journey begins, despite endless countermeasures (like hyperbaric chambers to prevent muscle loss) devised to alleviate this. Never before seen health issues will unfold as the journey progresses – physical, physiological, and psychological, and that's just in the eight month trip to Mars.

For the rest of their lives the travelers will be breathing artificial air, consisting mostly of argon, on a regimen of vitamins and other supplements, and will never eat regular food again. All this in an effort to lift the nations spirits and bolster our sense of pride? Bravo!

But the FDA doesn't test or regulate vitamins – nor does the FDA even approve their use! And they're *supplements* – intended to *supplement* an already healthy diet. There are questions regarding their ingredients and quality control, even their usefulness. Is the health and well being of our national heroes – no, the future of mankind – resting on the completely untested theory that man can live on vitamin supplements?

WHICH BRINGS US CLOSER TO THE MAIN QUESTION. IF YOU THINK the ultimate goal is to transport us and our culture and state-of-the-art technologies to a comfortable Earth substitute, think again. It seems that this could not be about preserving life as we know it – modern day man – certainly not the self-actualized, lifestyle obsessed 21st century man. Such a thing could never be maintained on Mars, not for a minute. But it seems that the other extreme, with man preserved in some primitive, unrecognizable form – would defeat the purpose. Does it even really matter what form of life? Plant, animal, microbe – anything with DNA – why would it matter? Life itself is such an inscrutable miracle, let's just get something up there. But it seems our vanity will not be denied – it needs to be us.

So, this might be one real possibility, in the form of a compromise: imagine a low-tech, machine-less existence – as before the Industrial Revolution. The Stone Ages.

Machines require maintenance and maintenance facilities, and early Martians will simply not have the means to support even the simplest machine-based existence. Academics joke about entropy but on Mars we'll see it happen right before our eyes: the machines and technologies we might bring would wear out and break down, and the Shuttle or SpaceX will not be coming to the rescue.

> **It will be a steady, downhill slide. A single generation of Martians, perhaps, will experience all the ages of man, from Modern Man, down through the Middle Ages, then the Stone Ages, ultimately living in caves and using rock tools.**

That might actually be the goal from the outset, at least partly – since so-called lava tubes have been detected on Mars, which are cave-like structures created from flowing lava – which is how they're created on

Earth but Mars also has volcanoes, though dormant. It's infinitely more feasible to maintain things in survival mode, as repugnant or dystopic as it might seem, rather than some sleek modernist dream or anything that is in the least bit in sync with Earth as many futurists or fanatics (or future SpaceX brochures) would prefer or fully expect. As they say, you can't have your cake and eat it too. Unless of course you expect cargo freighters to be landing every day from Earth, which would NOT be happening, for reasons the least of which includes the unsolved and vexatious problem of landing large payloads gently. Such a stone age existence, starting out, would severely limit Mars-based science (which is just more Earth-brain thinking), but it seems we're doing much of that already, with the rovers and similar remote means – which may as well continue as a jobs program if nothing else. So much effort will be spent by the early colonists to avoid the decline to some retrograde state, it may make sense to just start out in that retrograde state, including false teeth (since false teeth are easier to maintain than real teeth and you don't need teeth to eat flavored paste), no appendix, and other modifications or "surgical adaptations" to avoid the problems of maintaining such things.

So get ready for your new stone age existence on Mars! – truly a dream come true for any adventure seeker. The act of genius will be to recruit people for this!

LOW-TECH NATURE OF COLONY FOR FIRST 100 YEARS

The first colony, or even looking ahead a few hundred years, can't be both high-tech and self-sustaining, these two characteristics will be at odds with each other on Mars. It's just not a realistic expectation, is deliriously hopeful and just plain ridiculous. Because with high-tech you get more problems with things not working, malfunctioning, wearing out, and there won't be the inventory of replacement parts or repairman skills. You'd need a lot of jack-of-all-trade handy-man savvy types, brandishing skills with *MacGyveresque flair*.

The first generation of technology must be unbreakable, foolproof, and failsafe – because you don't want the underpopulated skeleton-crew first colony to be running around with repair manuals, spec sheets, and pocket-size 3D-printers, the interplanetary deus ex machina of the 21st century! That would not be getting off to a good start!

This obviously implies a more likely scenario: a first generation of technology that is simple and low maintenance. In science-fiction, the plot sometimes turns around the discovery of some technology – laying dormant for eons, built by some super intelligent species before they became extinct – as if in anticipation of becoming extinct and who then manage to go extinct – which magically turns on at just the right time for the star trekkers who "bumped into it" – and it works perfectly (in the manner intended by the aliens who built it if not for the benefit of the klutz who bumped into it).

That's what we need to be already on Mars when the first humans get there – a fully functioning habitat that is invincible and comes with an ironclad 1,000 year warranty – otherwise you'll have the hapless, asphyxiating space junkies who just wanted to go on an adventure instead running around with spec sheets.

This exact plot device is demonstrated in at least two sci-fi movies, including "Total Recall" and "Mission To Mars" – both set, interestingly enough – on Mars. Hopefully the Mars advocates are not inspired by either of these far-fetched and silly fantasies, the former of which even includes a Hollywood version of terraforming – ridiculous to say the least, junk science at its best.

So, the first colonists will not be making things with plastics, ceramics, and all the other wonderful things we have here on Earth (despite what some Mars advocates have actually predicted) – such ridiculous luxuries! Their time and energies will be consumed by the much simpler priorities of staying alive – and procreating. Adventure?

Farming may be the primary daily activity – which is, of course, hard work. We are going to Mars – to become farmers!

Think "Little House on the Prairie" – the antithesis of adventure, actually. This doesn't just get close to the idea of a prison planet – we've zoomed right past it through to the other side.

And the mothers will stay home to tend to the young (who will be in various states of adaptation and maladaptation, some cute, some not) and manually grind whatever will grow on a crater's surface into something edible that will keep them alive – and with a dash of artificial flavors from Earth – tender mercies! There may be few consolations that can be shipped via rocket from Earth but at least they should have that!

This will be the "life of adventure" on Mars promised by SpaceX – so spin that dial-a-flavor like a roulette wheel for the day's big surprise!

A deus ex machina, in case this isn't clear, is used by story tellers to rescue the protagonist but in a way that's not at all consistent with the storyline – imagine the hand of god coming out of nowhere to rescue the protagonist just when he's given up hope – it makes for a happy ending, but is completely illogical. This is what the Mars advocates are expecting, if unconsciously, when they imagine man colonizing Mars – that there will be some as yet unknown deus ex machina, in the form of technological breakthroughs that will "just happen," and all the critical problems will suddenly just be gone. We'll just schedule a rocket launch for some arbitrary point in the future, say a nice, safe 20 years out, or whatever – which SpaceX has done – and all the necessary breakthroughs will "just happen" in the meantime.

Given the unlikelihood of survival of a colony that is anything more than rudimentary, it seems that to even make a case for manufacturing on Mars is painfully unrealistic. Throwing 3D-printer technology into the equation – as some have suggested – does not seem to make a meaningful difference. In fact, it seems that any vision of manufacturing on Mars may miss the point entirely of being on Mars.

If Mars was a rest stop on a much longer journey to an actual Earth equivalent (or something better), there might be some point in establishing a colony on Mars. But surely this is just more science-fiction. Mars could never serve as a permanent residence for human beings who expect to have any degree of culture or Earth-like existence – certainly only with extremely down-scaled expectations could a colony have any chance of working. The more primitive the settlement, the greater the potential for even minimal success on Mars – but this certainly precludes a colony as enormous as the 80,000 proposed by Elon Musk of SpaceX.

On Mars we'd be starting from zero and never get beyond the equivalent of Earth's Middle-Ages – hopefully without the same myths and superstitions – while retaining perhaps a few modern advantages afforded by our knowledge of science generally, regarding nutrition and hygiene. But most modern contrivances and ideas would turn back into science-fiction – and become the stuff of legends.

43

Phase One Scenarios

The metaphysical "what am I doing here" will become one of man's first questions after he gets to Mars, and "what was I thinking?"

I suspect we will, of course, have already sent plant and animal species to see how they interact with the Martian laws of nature before sending people there.

ON EARTH, EVEN WITH THE MEANS TO TRANSPORT WATER OR FOR man to get to water, man still lives close to water. Lacking the means to get to or transport water on Mars, this proximity to water would be even more critical on Mars, where man would need to live in the immediate vicinity of any water source. This means one thing: anybody on Mars would have to live at the north pole – on the polar ice cap – with its extreme cold temps, since that is the only known location of an "accessible" water source, though in the highly inaccessible form of super cold ice. Mars' south pole ice cap is covered by a layer of carbon dioxide ice year round and whatever water ice is therefore *never* accessible.

SIZE OF FIRST COLONY

Three to four people wouldn't be enough to start the first colony, there'd be way too much work to do for this few people to handle by themselves. You'd need guys setting up the hydroponic equipment, other guys to maintain or monitor life support systems, at least 1 guy to monitor the air pressure and oxygen levels of the habitat/perform housekeeping duties while the others were out "exploring," and so on and so on. All this would exceed the brainpower, physical abilities, and stamina of just three or four people, all encumbered by spacesuits and reduced gravity (among other things like hunger, fatigue, etc.). All this would be way beyond the capacity of just 3-4 guys *even if everything was working perfectly!* The trip itself, including the take off and landing, would probably end up killing some of the crew members – even "slight injury or illness" during this could have far-reaching consequences. So there would need to be at least 10 or so in the first crew sent to Mars, plus a few more for redundancy or to even out the "group dynamics." In which case 10 might be the minimal crew size, 15-20 even better to really guarantee any kind of success with an initial colony if it's expected to be the beginning of something long term and self-sustaining.

BUT WHY MUST OUR EXPECTATIONS BE SO MODEST?

In setting up a "bare bones" colony we would certainly be setting ourselves up for failure! Must man's future be compromised? Considering that additions to an initial colony wouldn't occur for at least two more years, that there would be illness, injury, and death during this time (not to mention a waning sense of enthusiasm or other "monsters of the mind" among at least a few), and that some division of labor would be necessary, an initial colony of 100 may be even more beneficial, including women, to "round things out" and have some semblance to "life as we know it." So the size of the first colony should be at least 20 and preferably 100. The more the better.

After all, this isn't some trivial Boy Scout experiment! – we're talking about the future of mankind! Why would anyone even suggest sending up just 3-4 guys if a colony on Mars is in fact "vital?" – as the Mars advocates claim. Because if we send up only a few guys it seems this will be just another case of "been there done that" – and that that was the intention all along. And just a geeky experiment, like with the Moon. So let's build launching platforms in every country! An entire space industry on every continent! We don't just need one Elon Musk, we need 100 of them! It would certainly take nothing less than this to fulfill our destiny in space! – starting with a self-sustaining colony on

Mars – as opposed to a mere "jobs program" employing a couple ten thousand. Why aren't these things already happening? If it's vital to the future of mankind, *and the future of mankind is obviously vital!* – why isn't there already a GASA, a global version of NASA?

Two rockets preliminary to manned rocket

There will need to be a sizable inventory of parts and modular sub-assemblies on any rocket going to Mars, for instant snap in and snap out. There may need to be several rockets going up at once, some day in the future, some manned, others unmanned, to carry sufficient people, supplies, and equipment needed to provide some guarantee that a colony – no matter how rudimentary – can at least get started.

It's ironic that I take the view that it can't be done, then suggest how it might be done. It would certainly be convenient if there were accommodations waiting for Mars' first earthly visitors, so let's picture a scenario, one that would seem to be within our means: NASA sends up not one but two unmanned rockets, launched during the same window, each with the same payload, which would include a module for sleeping at least five people to be ready and waiting for the first residents when they arrive two or more years later. Let's assume the exact landing spot can't be predefined, down to the meter. The idea would be for both rockets to land within an area of a 2 mile radius. One of the rockets will be redundant, but since a launch window only occurs every 26 months, it's inconvenient to wait another whole 26 months if only 1 rocket is launched but fails to get there intact. So let's say both of the rockets land safely, within an area of a 2 mile radius. That gives the next rocket, 26 months later, the one carrying the astronauts, some wiggle room regarding where it lands. Because you don't want the astronauts landing 5 miles or more from either of the first two rockets. Because of the risks involved in getting to either prefab base camp this distance needs to be kept to a minimum. Even if the first two rockets land at the furthest points of the landing area, that's 4 miles apart. As long as the rocket(s) with the astronauts lands anywhere within this target area, they'll have to hike, at most, just over 2 miles to either base camp, with a good chance of this distance being a mile or less. Even if it lands outside this target area, having two base camps from which the first astronauts can choose may provide a "fail-safe" opportunity for success that may not otherwise exist. If the astronauts have a motorized vehicle in which to drive from the landing site to the base camp, a larger target area could be defined.

The manned rockets that went to the Moon carried no cargo, just Neil and two others and the shirts on their back because there was no intention to stay or set up a colony. Neil Armstrong and Buzz Aldrin were only on the Moon for two hours. And those rockets were huge – nothing but a giant fuel tank. So imagine how big the rockets would need to be if you're taking all your stuff! And enough food for at least two years, and equipment stolen from the intensive care unit of the local hospital to keep you alive on Mars, and your down comforter (with the air sucked out of course). It stands to reason there would need to be many unmanned rockets going up in advance of the astronauts. Perhaps one rocket per ton of equipment? There's no limit to how much stuff there would need to be already up there when the first manned ships got there if comfort and convenience are a concern – and they are. This equates to many unmanned rockets, preferable to manned rockets because they're much less complicated – although this preference may change because the idea is, after all, to get people to Mars and distributing many astronauts among many rockets may be less risky overall than on fewer rockets. It seems that either way, there'd be a lot of nail-biting going on – and that would never end. It seems that "the Mars mission" would become an ongoing preoccupation and source of anxiety for many people, call it "Mars anxiety."

About that first "giant leap for mankind," where Neil Armstrong and Buzz Aldrin were on the Moon for 2.5 hours: that was the plan, their visit wasn't cut short due to unforeseen circumstances. In a 2010 interview, Armstrong explained his moon walk might have been longer were it not for "concerns about the spacesuits keeping them warm" in the Moon's frigid temps. Do you see the "engineering" mentality at work? It's amazing how we rationalize to protect our vested interests. All five subsequent Apollo moon walks were only a few hours longer – were they still uncertain about the spacesuit? It seems concerns about the spacesuit would be the least of their worries – the *trusty spacesuit*. Rather, the moon walks were all so limited in scope because – can we be honest? – *there are no Holiday Inns on the Moon!* And because humans...belong on Earth.

> **Maybe we're glad that all the presuppositions about a lunar colony are behind us and don't really want more of it.**

So you're landing on Mars. It may be necessary to orbit for a while, and reconnoiter, in case the landing site is windy, it's night and you've planned a daytime landing, or to perform last minute calculations. Mars has days and nights similar to Earth (with a day of about 23.5 hours). The temperature will drop at least 50 degrees F in the hour or two after sunset, another 30 degrees or more before sunrise, and this generally will be the daily routine. From a high of around the freezing point (of water) or below during the day, to around minus 70 degrees F or colder, depending on where you land. The nights will always be pitch black because there is no moon – spooky. At least no moons like the one we know on Earth.

> **Mars actually has two asteroid-sized moons, Phobos and Deimos, which mean, appropriately enough, fear and panic.**
>
> **Things will look different when you're actually on Mars. Let's hope the astronauts can convert their enthusiasm for "going to Mars" to enthusiasm for "being on Mars."**

A prefab base camp might provide something similar to or *smaller* than a tiny studio apartment, and consist of a single or simple set of interconnected units. Despite the painful lack of options regarding this starting out, this arrangement would allow for basic communication and a social fabric, otherwise hindered by the constant and time-consuming suiting up and down to get from one unit to another.

Even in a Martian community of a few hundred it would be a cruelly unfulfilling existence. There would be severe limitations regarding human activities – hobbies; recreational, pleasurable and leisure activities; artistic and other creative interests – about which the human spirit has shown itself to be extremely demanding. Modern man does not have a survival based mentality, nor one that is accepting of any form of repression or oppression – these are in flagrant violation of our Western, democratic ideals, what we associate with "third world" or "developing" countries and to which we stick up our noses. We have evolved with the focus shifted from mere survival to an existence filled with fun, convenience, and "self-fulfillment" – and very safe but to the point of being unconscious of this – effortless.

> **This is Modern Man – smooth, effortless, elegant. We certainly can't be preserving this on Mars, where the focus will be shifted back – way back – to mere survival and staying alive, with NONE of the things we associate with Modern Man.**

There wouldn't be any culture for the first settlers. No fun. No color. The food wouldn't be appetizing, squeezed out of tubes. You wouldn't be able to call your friends back home because of the 6-42 minute lag, which would make two-way communication impossible. Which means you would not be able to interact with people back on Earth – ever. If you said Hi to someone on Earth, the signal would take at least 3 minutes for them to receive it, and another 3 minutes for you to hear their response – that's when Earth and Mars are at their closest. This 6 minutes would become 42 when the two planets are at their furthest, at opposite sides of the Sun. This also means it would take as much as 21 minutes for an internet page to load on Mars, unless of course your favorite pages are loaded ahead of time (see The Internet chapter). The first casualties will be death due to starvation or malnutrition, or suicide, some having eventually gone mad – who would have thought that home-sickness could be fatal?

Sometime during the first two years (and as early as the second week), which is arbitrary but an exact time is impossible to predict, something akin to the 7-year itch will occur. Some will have signed up not really knowing what they were getting into, especially if the space program isn't brutally honest about it and instead emphasizes the propagandistic "riches and heroism" as Robert Zubrin does in *The Case for Mars*. As the months pass, those who've been there the longest will have noticed how slowly things are developing. There will have been numerous delays, cost overruns, accidents, some – probably many – will have died due to failures in the life support systems. The primitive state of things will be persistent, and the point of it all will grow dim and dimmer. They'll get homesick. Life on Mars will be more stressful than anticipated. Many will become disenchanted with the whole thing, the thrill will be gone, it's just all work and no play. One decides he misses baseball, another his family, another decides he just wants to go fishing, many will have ongoing anxieties about the life support systems, some of which will have proven unreliable or just don't work as expected – which means people will be dying. It's easy to imagine a catastrophic failure in one of the systems that will cause either instant death by itself or through some domino effect or other "black swan" event. And they haven't even started the terraforming yet! Which, from the standpoint of those on Mars, will seem more dangerous and less worthwhile – maybe a little crazy, time to rethink the whole thing. By the two year mark all the members of the colony may have long given up any hopes of success.

The problem with low morale will become serious, and combined with a renewed and profound sense of appreciation for the big blue marble and all the opportunities and wonderful things that they (or their kids) will never have on Mars, they will all want to return to Earth! Every time they watch a movie they'll wish they were back on Earth. They'll have urges to do things that they've never done before on Earth but that can't be done on Mars. Even with wars and global warming, Earth is still better than Mars! Plus some will want kids and they don't want them to be Martian because all the other Martian babies are, well, strange, plus there's a high mortality among the newborns – no one knows why! – which is demoralizing. Even if just a few demand to return to Earth that would put the whole colony in jeopardy. And if their Earth-based counterparts don't satisfy this demand they'll become a threat to the entire project.

MOB PSYCHOLOGY

When the colony is small it's easier for everyone to "think alike," to be on the same page. But as the colony grows, it will become more diverse. Personalities will emerge, leaders and followers. Some will have more of the "adventure seeker" agenda – which will conflict with those having a more scientific agenda. Will these mixed priorities of the colony foster life – or death? Some will be more "by the book," favoring Earth-based rules, others will want to take a more heuristic approach to solving life's little problems – as they discover that the instruction manuals become more and more useless, full of naïve, Earth-based assumptions – after all, how could anyone on Earth know what Mars would be like? The colony may take on a direction completely unexpected by the Mars advocates – and will, in every way, be beyond the reach of NASA or the Mars advocates.

Synchronized cultures?

Some think that a Martian colony would somehow be an extension of America, but if man ever gets to Mars there would be absolutely no parallels to the way he lives on Earth, regardless of when, or the scale or scope of any colony. Instead of assuming that Earth and Mars will be somehow synchronized, or that each will draw from the other in some beneficial way, we should assume the opposite. Martian culture would diverge sharply from Earth culture with, for example, the Martians inventing new words and changing the language, so that Earth and Mars no longer understand each other, can't relate to each other, where the need to communicate with each other becomes less and less, the routines and customs of each becoming irrelevant to the other, such that each considers the other to be alien. *Alien.*

NASA and other entities will design plans by which the first settlers should be governed, but at some point these will be out the window. Life on Mars would be determined by Martians, not legislators back on Earth, bureaucrats at NASA, or other preconceived style guide. This would add to the culture shock of subsequent waves of settlers as they arrive with the NASA rule book in their back pockets.

Each successive generation of Martians will be less earthly and more Martian, so that future waves of settlers arriving from Earth would be treated as unwelcome, outsiders, like illegal aliens

in a way that would sabotage the ongoing emigration of earthlings. Future generations of Martians may not be as accepting of "ethnic diversity" as we are on Earth. Add to this the generation gap that would inevitably exist between successive generations of those born on Mars (as on Earth), compounded by genotypic variations as the Martian species evolves, generating animosities and rivalries that could destroy the long term prospects of a permanent colony.

Imagine putting 8 or so Americans in some desert, the Sahara, for example, where there was nothing. Then in two years add 8 more. Those already there would have learned to do things a certain way, which could be vastly different from how they were initially trained. The problem caused by the new group clinging to their training would be minor compared to the larger problem of balancing the responsibilities among all 16 that none of the 16 had expected, and with each new group of 8 (or however many) this rebalancing would need to occur. You wouldn't simply have more people doing the same things. Like when someone is born, the dynamic of the family changes, only with the growing colony in the desert the shifting dynamic would be more dramatic. Now imagine this desert colony, that has every advantage of Earth...on Mars. How different would the first group of 8 (or whatever) colonists be after the first two years when the second group of 8 arrived? They would have become bona fide Martians. Would they welcome the next group of earthlings with open arms? Each successive group of 8 or so earthlings would pose a threat to the colony, in terms of carrying diseases from Earth, or be disruptive in other unexpected ways. After a few dozen or a hundred colonists, the dynamic may become too fragile, or unsustainable, such that continued growth would need to stop. Shifting population dynamics every 2 years as described, along with ongoing cultural and environmental adjustments addressed elsewhere in this book, would add to the complexity of things, and this complication factor – as with any complicating factor – would lower the odds for the success of a Martian colony. It would be easier at first, for the first colonists, to be selected based on compatibility with each other, but this will become impossible with each successive wave of new colonists – every colonist won't be compatible with every other colonist – and negative personality traits will inevitably emerge and put the colony at risk.

Just packing boxes is strangely fatiguing, as I learned
recently before moving cross country. My eyes felt like they were drying
out after a while of scrutinizing things, examining everything I own to
decide whether to bring in the car, ship, store, sell, throw out, give to
charity, give to the next owner, or give to friend or family. Then of those
items being boxed, to determine the number and size of boxes, since
there is some ideal combination of boxes that would allow me to maximize what I could carry in the car and to minimize what I would ship.
It's not just a simple matter of getting x number of same size boxes.
One ideal solution is to get several medium boxes, around which one fits
smaller boxes of varying sizes, which fit better into the curved areas of
the car trunk, so I learned to put smaller things in the smaller boxes. (I
constructed all the boxes first and put in the car while still empty, testing
various size boxes for arrangement and fit.) Plus I wanted to stack boxes
in the back seat, but not so much that I couldn't see out the rear-view
mirror comfortably or the back side windows, and even the space between those boxes and the rear of the front seats, in which I could fit
narrow or unboxed things. I learned to put larger things in a box first,
then pack smaller things around it. So you pack things according to
size, rather than by the rule to pack things the way they might already
be grouped in the house or apartment. Am I boring you? That's my
point, to evoke the ennui that will be Mars. Could one be more careful
in packing – a rocket to Mars? In this example it's ultimately arbitrary
what goes into what box, no decision being a matter of right or wrong,
life or death, but I imagine living on Mars would be full of such mundane
decision making – and nonstop.

> **Imagine the fatigue that must develop, over time –
> but where no decision is arbitrary, always careful,
> where it's not just a matter of whether some decision
> will affect mine or another's life (BECAUSE THEY
> ALL WILL!) but will, in addition, be a matter of some
> carefully calculated compromise based on numerous
> critical variables – and compounded by the opinions
> of others in the team or colony.**

Is there no compassion? What torture! Imagine if all you did was
pack boxes, then unpack them, and that was your life. Pack, unpack,

pack, unpack, or some other similarly monotonous routine. This would describe life on Mars for the first generation, at least.

Scuba diving is a familiar example of an activity that requires a lot of attention to detail while at the same time involving some real danger. At least you know what you're getting into even before your first dive. You already understand what water is, and how to swim, and you've seen others doing this, and what things look like underwater, so you know even before your first dive what it would be like – or that you would even want to do this. Now, *without* any of the foreknowledge provided by this, would you agree to...live under a dome at the bottom of the deepest sea with no chance of returning to land, no ability to speak to anyone on land, no possibility of rescue from anyone on land, unreliable sources of food, oxygen, and water, *and* had to wear a stifling spacesuit unless you were inside an interior habitat, etc., etc? Of course not! But this is what living on Mars would be like, in a sense, being underwater on a 24/7/365 basis. And likewise you would have no chance of returning to Earth if you decided you didn't like Mars.

How THINGS EVOLVE ON MARS – THE ACTUAL COURSE OF SUCCESS or disaster – can't possibly be imagined with our Earth-centric brains. The degree of development or sophistication will be determined by Mars, as will the extent to which the residents of Mars can develop the tools to do this, and other Martian priorities – not by what we on Earth think Mars should be. All this will be determined by the constraints imposed on man by the Martian environment and the Martian laws of nature, which we know are nothing like that on Earth. We earthlings can't impose our will, or our expectations, on Mars, or on future residents of Mars, any more than we could impose our will on any other alien intelligences in the universe. Humans: the cosmic control freaks.

44

Technology & 3D-Printing

Martians with 3D-printers may be no different than cavemen with 3D-printers.

THE STATE OF THE ART ON MARS, REGARDING ALL THINGS — AT best – would lag at least two years behind that of Earth (which corresponds to the time between launch windows, which determines when and how often rockets can be launched from Earth to Mars), and this lag would increase over time, given the increasing self-sufficiency that would be inevitable due to the absurd cost of staying in sync with Earth – as if that was even an assumption. The Martians would be living in a different time frame altogether. They would not really have the means to build or make things we have on Earth even if they had the blueprints – because it seems the Martian brain would have a completely different set of priorities than its earthly counterpart.

On Mars we wouldn't be able to just phone in an order for that new cancer drug, that paradigm would be utterly, completely gone. We wouldn't have the basic R&D of even a third world country. We could learn of the latest tech but would always be limited to what could fit in the cargo bay of the next rocket ship that would take at least 7-8 months to arrive – no overnight deliveries to Mars.

It might actually benefit future Martians to be ambivalent about Earth, the two cultures would be that different.

3D-Printing

One interesting thing about modern man is our reliance – or over-reliance – upon electronic technology, our expectations of it – which are often unrealistic, as if every new gizmo really was something great and wonderful – and the urgent belief that the key to living on Mars will be founded on these technologies! That life on Mars may not be possible "quite yet" but for a few more breakthroughs that are just over the horizon because *we all know that breakthroughs have a way of happening just when you want them.* Some of our newer technologies, such as 3D-printing, seem to have an immediate application – and in the most critical way! – for living on Mars! Remember when we all ran out to buy color printers? What a fad that was! Now it's in 3D! And it will change everything! The truth is, 3D-printing *is* somewhat of a game changer – but can this translate to Mars? The scientific community – and science writers – are buzzing.

3D-printing technology, an innovation of the 21st century, would make an interesting convergence with any future manned trip to Mars, providing a technological boon that is decidedly unearthly but on Mars, low-tech solutions, as previously indicated, may be more feasible than hi-tech solutions. There would be no natural supply of oxygen, food, or liquid water on Mars, but...3D-printers? At first glance, 3D-printers might seem to be on the positive side, but that may not be saying much when the negative side is weighted so heavily – although NASA is exploring ways to use this technology in space. The payoff could be significant and alter the nature of the payloads, alleviating the need for an excessive inventory of spare parts or supplies. (According to one rule of thumb, in 2012, it costs $10,000 to bring one pound of equipment into space.) This would require that one go through the entire line of inventory and decide which parts can or can't be 3D-printed, which would in turn require the CAD (computer automated design) specs for each part of each mechanical device – which lies somewhere between daunting and impossible.

Some have gone so far as to suggest that a 3D-printer could create, using regular dirt as the substrate, any "fine-machined" part. Such an innovation would be earthshaking! We would no longer need chemical and material engineers to make us the stuff of our lives!

One could, of course, imagine the 3D-printer factory in a box as a dream machine, the sine qua non of man's Martian future. On the other hand, this technology may not lend itself to making reliable replacement parts for broken or worn equipment as readily as one might hope, and whether or not it can replicate Original Equipment Manufacturer (OEM) standards may be crucial, aside from some of the more fundamental issues concerning their operation and performance on Mars. The technology could, perhaps, provide an extra measure of *luck* under some circumstances. It might allow for more sophisticated jerry-rigging of replacement parts when things break down, such as in this example from Businessweek, Jan 2012: If Apollo 13 had been equipped with a 3D-printer, fixing the system that removed carbon dioxide from the spacecraft might have been much different. As depicted in Ron Howard's Oscar-winning movie, engineers at ground control used materials they knew were on the craft to jerry-rig a solution that they then shared with the astronauts. With 3D-printing, that same problem might be fixed by sending a 3D design to the spacecraft. "Instead of 'Houston, we have a problem,' it might have been 'Houston, we have a problem, so now send up a design so we can print out something here to fix it,'" according to Gonzalo Martinez, director of strategic research at Autodesk at the time.

But of course future residents of Mars or more distant locations would not be calling Houston – or anyone else on Earth – for help and would (presumably) already have all the relevant knowledge and "computer files."

What if your Nook or Kindle e-book reader malfunctions on Mars? It's gone. You can't just order another, nor will a 3D-printer be able to just "print" a new set of circuit boards, nor can you just bring up x number of spare Nooks despite their compact size. It's these small, unplanned inconveniences – or what would be considered small on Earth but would become HUGE when reading and other electronic based media and entertainment becomes a major part of your life on Mars, even for those who hadn't spent a lot of time reading as earthlings – or those living a life of adventure! Venture forth, adventure-crazed earthlings, and spend the rest of your lives on Mars...*reading*! Just hope your e-media devices outlast the "limited warranty" – and be warned there's no customer service on Mars! This will be the case with every single electronic device taken to Mars *and there is no customer service on Mars!*

> **Every single day the new Martians will think of things they wish they had brought and they can't just magically create these things with a 3D-printer or expect them on the next supply ship – because there will be no shipping service to Mars. The list of things will double in size each day that passes, along with the yearning and craving for those things, the totality of which will become a substantial inconvenience, an adversity, a hole burning in their heads. It will take a tremendous amount of will power, not just to adapt to the environment, but to the absence of comforts, conveniences, and amenities – and the hole burning in their heads. The early Martians will have almost nothing that they had on Earth – not even air. If this represents the new age of man...it seems it will not be a good start.**

So the prospects for 3D-printers on Mars may be less than we think. MacGyveresque ingenuity with low-tech objects on hand like duct tape and paper (along with hammer, screwdriver, and bare-knuckle practicality) will probably be more reliable than the far more complicated 3D-printer, which requires a power source, the right CAD files, and the right feedstock. In the Apollo 13 example above, use of a 3D-printer might have taken *longer*, and demonstrates that sometimes a quick and dirty improvisation is best, so we should suspect that a 3D-printer will not always be – may never be – the universal provider of instant solutions.

On Mars, low-tech solutions will often be more feasible than hi-tech solutions. On Mars, 3D-printers may never be a match for spare parts on hand or the ability to improvise.

> **Even if 3D-printed parts are equivalent to OEM parts, it may be a moot point. Using 3D-printer technology to make scale models, prototypes, and OEM parts is great on Earth, but getting things started on Mars will require far more fundamental innovations and leaps.**

Martians with 3D-printers may be no different than cavemen with 3D-printers.

45

Time Lag

‖‖‖

Real-time, seamless communication explains, in part, the dichotomy of manned/unmanned exploration.

THIS IS A LOFTY AMBITION, REGARDING THE 3D-PRINTING. THE preceding example (using Apollo 13) also illustrates the inextricable dependency astronauts have historically had on Earth-based technical guidance and control – Martian astronauts will not be afforded this luxury. This dependency can't continue on Martian missions because real-time, seamless communication with astronauts flying between Earth and Mars, and then with Mars, will not be possible because of the time it takes for communications signals to travel. There is a lag between the time a message is sent from Earth and received by the astronauts, which increases with increasing distance and – even when this lag is only a few seconds – makes meaningful conversations impossible. This was not a noticeable problem with the lunar missions, but Mars is much further away. Rockets bound for Mars will lose seamless communication with Earth just days into the 8-month journey.

This is a hideous inconvenience – the pain this causes NASA is one of its dirty little secrets, a nightmare – but it is a fact of life that can not be overcome, now or ever

since the speed of light is a constant and is what determines the travel time of the communication signals. This...inconvenience...makes travelling to Mars a completely different game from travelling to the Moon.

It takes between 3 and 21 minutes for a signal to travel from Earth to Mars, depending on their distance from each other. This translates into 6-42 minutes of silence after finishing a sentence and hearing the person's response – which, of course, makes talking on the phone – seamless communication – to Mars (or other planets) impossible.

Martian astronauts will have to carry with them, in their heads or meticulously tabbed reference e-books, not just the equivalent intelligence of dozens, even hundreds, of engineers and technicians, but the hands-on skills and tools necessary to execute rapid solutions or improvisations contrived on the spot, and without the benefit of models or testing platforms that would be available to Earth-based support specialists as with the lunar missions. Martian astronauts will not have this convenient accommodation, this shoulder to lean on – rather, this *vital link* – that was available to the lunar astronauts and that literally made the whole thing possible! At least not for matters of urgency.

**This is a daunting prospect, a bugaboo of space travel...
and the curse of all future interplanetary communication.
It's a problem never depicted in sci-fi movies, where one
merely suspends disbelief.**

The International Space Station (ISS) is in constant contact with Earth – which is of course key to its success. In fact it's only about 250 miles from Earth. There are only 3-6 people on board, but operation of the ISS depends on the teamwork of hundreds more on the ground – "teamwork" implying seamless, real-time communication among everyone involved.

Every space program so far involving manned missions into space has depended on the collective intelligence of hundreds of people, and which are synchronized in real-time, seamless communication. Lack of this collective intelligence, which would happen if you took away seamless communication – changes everything. This is why the dichotomy of manned/unmanned exploration exists, and which is of enormous relevance when debating whether man can colonize Mars. The reason this dichotomy exists is one of the reasons man can't colonize Mars. One might well ask why the two Voyager probes were unmanned?

Wouldn't they have been the ultimate adventure for some carefully chosen adventure seekers? Many of the same considerations of "man in space" apply to both the Voyager probes and colonizing Mars – so why do the answers seem clear even to the Mars advocates for one but not the other? Although we couldn't have sent men to Mars along with (or instead of) the rovers, we might have sent a man along for the ride in the Voyager mission, right? It seems the answer is simple: even with ample food supplies and creature comforts, he would be bored to death! So in that case the human element was, obviously, considered – and which prevailed. But the *human element* seems not to be a consideration in the case of colonizing Mars.

46

Generation 2

Oh, the glory, the wonder, of our beautiful, lovely water planet Earth! We will never run out of water, or air, or have to worry about air pressure or gravity or the freaking magnetosphere – at least not in my lifetime but do you think future Martians will be thinking ten generations ahead? Pity the pour souls on Mars in their struggles simply to survive and who will look upon Earth with weeping, envious eyes.

It may be inevitable that the long term goals of NASA and other space programs may be thwarted by...Martians!

HERE'S WHERE WE GET TO THE PART THAT SEEMS BEYOND THE grasp of the rocket scientists, architects of the space program, and enabling science writers – where the dream is most likely to fall apart like sand castles in a tsunami.

Can we assume that people born on Mars will blindly embrace the philosophy that got them there in the first place, and like mindless drones dutifully carry out their parents mission?

Or would they have the same feverish sense of curiosity and adventure as their parents and therefore think of Earth as being enormously intriguing? After all, Earth doesn't need ongoing terraforming, the success of which requires that many unproven assumptions work like magic – and which could backfire on Mars such that it becomes even more deadly than it already is. Either way, it would be a lot of work and have the colonists in a constant state of peril. One version of terraforming would employ ongoing fire bombing of the Martian landscape – ongoing in perpetuity even after settlements are built since *the efforts to terraform will obviously, if slowly, be undone by naturally occurring circumstances.* A great environment in which to raise a family, wouldn't you agree? Or for scientists to "do research?" The terraforming itself may be the instrument of its own destruction, the cause of a species-level extinction of all the colonists *along with whatever other new life forms.*

At best, some of those born on Mars might decide their presence on Mars to be nothing but a demented, deranged experiment, a form of enslavement. To future Martians, Earth may seem a lot more habitable than the planet they were born on! Their parents fears of the "long term future of Earth" will be rejected by them, while others "dream of a better life" and defect to Earth, perhaps as enchanted by the American Dream as everyone on Earth is. Escape to planet Earth! A planet literally covered with water! The Mars fanatics obviously do not anticipate this. It may be inevitable that the long term goals of NASA and other space programs may be thwarted in this way by...*Martians!* – who will of course have free will. Or will Martian teachers (Martian Illuminati) keep Earth a secret from them, create a mythology that Earth is some kind of death planet? And what kind of revolution do you expect when they discover that fraud? And how about when future Martians come back and mate with Earthlings? Will we love them *oh never mind!* The science, the ethics, and the legalities boggle the mind.

We're warm blooded! It's incongruous that we would live on a planet that is entirely frozen – and to think this is an inevitable choice or destiny.

Humans have been guided by evolutionary forces that have served us well, that have allowed us to achieve our current state of technology, and which allow us to live fairly comfortable and long lives. We trust these forces and exploit them to our benefit, rather than ignore or defy them. And being comfortable isn't a frivolous contrivance of the leisure class, it's a primitive instinct, since comfort equates to longevity. Even now, on Earth, we crowd into confined zones of habitability, in search of – obsessed with – comfort and convenience – and still complain – how precious we are – as if Mars would be better, a solution to overcrowding or overpopulation – or other modern problems.

> **In forming our ethics, we've somehow invented a "hardship is good" mentality, to rationalize the tough times, but it's outrageous to use this cliché as a cornerstone to defend a "mission to Mars."**

Given the scale of the Mars mission, and its impact not only on man but on Mars as well, if Mars isn't as conducive to human life or almost, it's foolish to think we can or should go there – and it has clearly been established that Mars is not as conducive to human life or almost. Life on Mars would be very heavily compromised, if not impossible, and not defensible with the trite "hardship is good." To subject future generations to such a misbegotten hardship requires a blinding degree of both arrogance and carelessness. Do you really want your children, or children's children, to live in a constant state of machine-based life support or grossly mutated? Or a primitive, dark existence free of ambition or hope?

> **An atmosphere devoid of oxygen isn't, in some twisted ethic, good. No water is not, in some weird, outer-space-logic way, good.**

Whether or not there exists scant amounts of oxygen or water is pointless argumentation and hopefully the most radical of the Mars advocate space cadets will some day appreciate this. Our future success in space can not rely on the jury rigging of natural phenomenon that are clearly antagonistic to us with dumb, science-fiction gimmicks like the so-called terraforming. Our past space programs and a few people on the International Space Station do not justify or support the belief that we can or should leave our friendly Earth ecosystem, tricked by the Arizona-esque illusion that some can't resist – and which could easily be resolved by simply moving to Arizona – it's surprisingly uncrowded!

IMAGINE THE FIRST MARTIANS COMING OF AGE. CAN WE EVEN assume a person born on Mars would be physically and mentally healthy, sufficient to become a well adjusted, fully functioning adult, capable of maintaining the Martian Earth substitute? Or by Earth standards will "fully functioning" on Mars be psychotic? Either way, they may see the same internet you and I do – they'll want what they see – and be stowed away on the first rocket ship back to Earth. Could this happen? Imagine the second generation of Mars born humans defecting to Earth to get jobs at Mars One to promote space tourism on Mars!

> **The entire second generation will return to Earth, leaving their idealistic parents to die off on Mars, and their dream with it. Therein lies one possible future and the quick death of the entire Mars mission. Getting there never really was the problem, nor the distance. The fundamental elements of man are not a matter of rocket science, nor the product of technology; they can't be engineered.**

So it seems inevitable that future Martians will return to Earth like modern day Cubans risking life and limb swimming to America or others who stow away in wheel wells of airplanes or cargo holds of ships – seeking (as if you couldn't have guessed) a better life!

> **Come on, tell me you didn't see this coming from a mile away! Why would a human born on Mars embrace the ideals of its square parents – since when has that been human nature? You'd have to expect a high rate of defection, as soon as the means became available, unless of course the Mars fanatics put up a Berlin Wall between Mars and Earth. Is this part of their plan?**

The idea that children will be born on Mars and grow up to be healthy and well adjusted scientists begs to be ridiculed, when one considers that they will lack the basic experiences of being human: bubble gum, flying kites, baseball, fishing, comic books, yo-yo's, Frisbees, skipping stones, skinny dipping, goldfish, blue jeans, ski sweaters, surf boards, rock 'n roll, penny loafers, hot dogs, apple pie, Chevrolet, soda pop, candy bars, roller coasters, bicycles, the county fair. Breathing air. But not just *air*: mountain air, spring-time air, the air at the sea shore, the air after a summer storm...and all the other things that give us a reason to have five senses! Who needs these things, really? They are all so silly! But if we're not preserving these things, what creepy Martian-human creature do we end up with?

CLONING

Thinking we can put men on Mars as some altruistic gesture to the future or to preserve our "legacy" – with the assumption that your Martian kids will embrace and put into practice your ideology – is like creating a clone of yourself with the intention of using it for spare parts while ignoring the fact that a clone of yourself is human and would have free will and dreams of its own. If you clone yourself, you'd be creating another human being who would, of course, have the same rights as you – you couldn't just put him in a box like some inanimate object while you wait for your heart or liver to fail, or attach him to feeding tubes and pretend he's not aware and sentient like yourself.

The Island demonstrates just how dystopic a future with human cloning might be. Set in the near future, this movie is about a society where cloning is used by the rich and powerful to live longer – the clones are used as a source for organ transplants – but only through a system of highly effective propaganda that hides the truth – which is that the clones die or are killed in the process. In this particular vision of the future, it's made clear that clones are people no different from you and me, and that exploiting clones for this purpose...is murder.

And only through a system of highly effective propaganda as in *The Island* could one possibly believe that man can live or flourish on Mars or that there could be any kind of permanent colony on Mars. That system of propaganda is in full force, through the Mars advocates, which of course includes NASA.

Men on Mars are going to have the same ordinary, mundane priorities that he has always had, which is to say he will not be particularly worried about "the future of Mankind." He will be concerned with HIS survival, NOT that of his species, and likewise with his own sense of satisfaction and personal fulfillment. Man is inherently self-centered and this will not be different if he is living on Mars.

47

Exoplanets

When an astronomer says "Earth-like" he doesn't mean that if you go there you'll think you're on Earth – in terms of the atmosphere, gravity, lakes, rivers, mountains, etc. – which is embarrassing because that's what I thought! Nor does he mean in the slightly broader sense of livable – but there is no broader sense!

EXOPLANETS ARE SIMPLY PLANETS IN SOLAR SYSTEMS BESIDES our own. Our solar system consists of the sun, around which orbit the 8 familiar planets. Pluto, which had been the ninth planet, is further out than the others and within a large zone of smaller orbiting bodies to which it is more similar, hence Pluto is now grouped with them rather than as a planet – fancy that.

There are many solar systems besides our own – *billions!* – which comprise our galaxy, the Milky Way. It's incomprehensibly enormous, so much so that we don't measure it in miles because the numbers would take too long to write, so we measure distances within the galaxy – the distances between the suns, or stars – in light-years. A light-year is simply the distance light travels in a year. Light travels fast, at about 186,000 miles per second, which in a year is 5.87 trillion miles, or about 5,874,000,000,000 miles – that's a light-year.

Planets in other solar systems are different from the ones in our solar system – all the solar systems are a little different, some a lot different. The Kepler satellite, launched in 2009 and which orbits the sun, is used to detect those planets – some of which have been declared Earth-like. This doesn't mean what most people would expect, actually supporting an Earth-like ecosystem, or even having that potential, but the contemporary definition of Earth-like in this context is overly simplistic and will be refined as our knowledge grows. NASA admitted that Kepler's performance as a telescope was not as hoped, and in August 2013 declared the mission more or less crippled due to certain mechanical glitches. Nonetheless, it has allowed scientists to *guesstimate* that 17 billion Earth-sized exoplanets reside in our, the Milky Way Galaxy [Space.com].

And just think, even if there were Holiday Inns flashing "Welcome Humans" in red, white, and blue, no human will ever travel to any of them!

They're all light-years away – every single one of them, even the closest ones, so spare yourself the number crunching, all you data-crazy star gazers. What's mind-blowing isn't the "17 billion" but the irrational exuberance surrounding it. In this context, our ability to collect and accumulate data and metadata has clearly surpassed our ability to use it. Earth-sized, Earth-like – the terms are apparently used interchangeably, in an effort to make the cosmos seem less alien...by those who need to extract some comfort on those lonely nights in the observatory. As if anyone lost in space should head for the nearest Earth-like planet!

The closest solar system to ours is Alpha Centauri. It's 4.2 light-years away, meaning if you turned on your flashlight somebody standing on that star would see it in 4.2 years, with the beam of light travelling at 670 million miles an hour. It's actually a system of 3 stars, which might have a planet but not in the habitable zone. This is followed by a star named Tau Ceti at 12 light-years (considered to have 5 planets, one possibly in the habitable zone but very prone to impact events). Star Gliese 581 is also out there, 20 light-years away.

So if you could travel at the speed of light to Alpha Centauri, you'd be there in 4.2 years. It's interesting that I would use this analogy, rather than the flashlight analogy, because shining a flashlight out into space is easy to understand, and is completely realistic, whereas a person travelling at the speed of light is not realistic, so it's funny that I would resort to this analogy when I'm presuming to be absolutely practical. Maybe you prefer this unrealistic and impractical analogy – it does have a way of capturing the imagination, so for the sake of convention, let's use it!

But if you were travelling to Alpha Centauri in a regular slow-poke rocket ship, the kind that NASA has used for years (and then private companies like SpaceX), it would take you MANY years to get there, about 40,000 – you'd be dead by the time you got there, which means we can't travel to even the next solar system over or its imagined planets, much less those that are even further away. In science-fiction stories the characters can do this, travelling among the stars very conveniently – but that's fiction. We live in reality.

Do you see the logistical issues involved with travelling to even the closest solar system? It would take over 4 years even if we could travel at the speed of light, for which we'd first have to be converted into pure energy, which of course will never happen. Speed of light travel is generally considered to be impossible even in theory because it would take more than all the energy in the universe to power an engine for this, according to experts.

For comparison, it would take the unmanned probe Voyager 1 over 374,000 years to get to Gliese 581 travelling at its cruising speed of 35,790 miles/hour, compared to 20 years if it were travelling at the speed of light. Do you think "aliens" wouldn't have these same logistical issues in travelling through space? The odds of alien species visiting Earth in our lifetime = the odds of us bumping into a dinosaur, and for similar reasons regarding the staggering vastness of both time and space.

The distinction made by *exoplanet* is purely for convenience, which begs the question, convenient for who? Because it all seems somewhat ridiculous to think that anyone would speak of planets in *other* solar systems. When was the last time you spoke of a planet that wasn't one of the familiar eight? As if we need to be familiar even with them! Even in science-fiction, when a story takes place on a planet or planets in other solar systems, these other planets are usually referred to in some scientific sense but without the excess, flakey rationalizing created by the label *exoplanet*.

Where things get really flakey is when astronomers start classifying these exoplanets as "Earth-like" or "Super-Earth like," in the completely new arena of astronomy called Goldilocks planets. Did you hear me? I said GOLDILOCKS PLANETS!

First of all, it's interesting that astronomers would devise a special category for Earth-like planets only now, as if one couldn't have assumed all along that every star has a similar arrangement of planets like our own, and inhabited. Yet still different enough to make the respective ecosystems incompatible.

One needs to be indignant with writers or scientists – the entire astronomical community – for using the term "Earth-like" in describing Mars, and various other planets, as if the comparison is intended to be literal – which it isn't. Or to mean that man can therefore live there. Or to imply life, possibly intelligent, is already there – which is absurd – the extent of the absurdity that is not already established scientifically can pretty well be surmised – in this arena things may not be as easy to prove or disprove as we're accustomed to, so certain leaps may be necessary.

Earth-like. What does this mean if not that humans can live there unencumbered by memories of Earth?

The justification behind so-called "Earth-like" planets is very sketchy, often having no Earth-like characteristics, and are said to exist in the so-called Goldilocks zone, a ridiculous contrivance, in this emerging vernacular, where science literally veers off into, not just fantasy, but childish fantasy. As if by using the term one fully intends to *not* be taken seriously. Goldilocks obviously has implications of warm and cozy, so it stands to reason that Earth is in a Goldilocks zone. But I would have thought the term would be confined to something even more specific, like my living room with the thermostat carefully set, which is – literally – warm and cozy rather than something that "may harbor life" but which may actually be the opposite, like planet COROT-7b, the lava-coated "Super-Earth" 489 light-years away. Lava-coated Super-Earth? Clearly "livable" isn't so much a spectrum as much as it is a very specific frequency. If Goldilocks zone is supposed to define that frequency, it seems to be painfully optimistic.

THIS CAN'T BE OVERSTATED. YOU SEE, WHEN AN ASTRONOMER says "Earth-like" he doesn't mean that if you go there you'll think you're on Earth – in terms of the atmosphere, gravity, lakes, rivers, mountains, etc. – which is embarrassing because that's what I thought! Nor does he mean in the slightly broader sense of livable – but there is no broader sense! He means in the astronomical sense of a celestial body that is the size of Earth give or take several orders of magnitude – and nothing else – because we can know little else due to their distance – and never will (except for perhaps those belonging to our closest neighbors like Alpha Centauri). The 17 billion isn't an actual count of 17 billion planets. It's a wild guess – an extrapolation – based on a long series of assumptions. Which means the number isn't based in science, but an atavistic yearning to *wonder*. Meaning there are not 17 billion other planets that we could live on. There are probably *none* – and Mars is no exception. This doesn't mean there can't be other planets that are so similar to Earth as to be an exact duplicate, just that we can never know this. The puzzle is actually more metaphysical than physical. Let there be ten Earth-like planets, or a gazillion – it wouldn't matter. So why is this so exciting to astronomers? – and not Saturn-like planets, or Venus-like? I want the counts for those! Crunch the numbers, dammit! Venus-like, Earth-like, what's the difference?

AS IF EARTH WAS OUR ADOPTIVE PARENT. SURELY THERE WAS some kind of cosmic mix-up and we need to get *home*. Along with calling planets Earth-like when they're not, it all seems so very unscientific, as if devised by an astronomy school drop-out, especially knowing that planets known to be uninhabitable...can still be classified as a Goldilocks planet! This is just not scientific.

The official designation "habitable," in the Goldilocks sense, relies on an extremely crude (unscientific) assumption, based solely on a planet's distance from the parent star. Nothing more.

But from this alone, scientists make the huge leap that this means any water on such a planet exists in a liquid state...and hence can support life as we know it. Because the technology does not exist (i.e., it's impossible) for us to directly observe or study planets in other solar systems, they're just too far away – just to conclude that the "wobble" of a star is caused by an orbiting planet is a resume building achievement. We can count the stars in the sky very easily because of their large size and high visibility, but not planets in other solar systems even with the most powerful telescopes – rather their presence must be inferred from other measurements (with a few rare exceptions that have been photographed directly like planet Fomalhaut b, 25 light-years away in the Piscis Austrinus constellation, using the Hubble space telescope) – but *none* can be studied until more powerful telescopes are designed. But what's the point if they're so far away? More powerful telescopes won't bring them closer. Surely, as with notions of speed of light travel, there is more inspiration from science-fiction than practicality.

Even the astronomers are confused. For example, the closest Goldilocks planet detected, rotating around the sun Gliese 581, 20 light-years away, has been described as the "most Earth-like," while the exoplanet GJ 1214b is described as "the Super-Earth closest to Earth," even though it's much more distant at 42 light-years away. Evidently an Earth-like planet doesn't become "super" because of characteristics we associate with habitability – as one might think – but because it's bigger than Earth, like Mu Arae c, a Super-Earth that is 900 degrees Kelvin on the surface – but which I would therefore expect to be abruptly disqualified by Goldilocks hunters.

In fact the International Astronomical Union has not yet defined an official system for designations of extrasolar planets, or exoplanets. The IAU only decided in 2013 to restart the discussion. With that in mind, one would think the designation "Earth-like" would be reserved exclusively for planets that have liquid water on the surface, and "Super-Earth" for planets with the additional feature of a living ecosystem comparable to Earth. But such a neat and tidy system could never be realized. The problem is, there is no way to detect these things on planets in other solar systems. Identifying the constituents of our own planets – and their moons – is difficult enough, is ongoing – and will never end. So it's understandable that the determination of life, or just the presence of liquid water – on planets in other solar systems isn't just impossible right now, but may always be impossible.

Even where things are relatively easy to study, like our own solar system, we tinker with dogma by reclassifying Pluto, such that it's no longer considered a planet. So further out, where things are more arbitrary, ongoing discoveries – enabled by increasing degrees of scrutiny – will defy our attempts to classify them with either certainty or consistency. For example three exoplanets are named a, b, and c in their order from the parent sun, but then another one is discovered closer to its sun. Calling it "d" would imply it's furthest from the sun – which would be incorrect – but renaming all the others to comply with the naming convention would confuse others already familiar with them. We could, of course, give exoplanets more meaningful names than a, b, or c but with thousands – millions – of exoplanets, how can each be meaningful? It's hard to believe that even one would be meaningful.

THE REASON IT'S ALL SO FLAKEY IS BECAUSE IT WOULD SEEM THAT the point in studying exoplanets is to find better alternatives to Earth than Mars, more suited to human habitation, more like Earth – exactly like Earth – BETTER than Earth!

If we could travel at some fraction of the speed of light we wouldn't waste a second even thinking of Mars.

Mars becomes "necessary" only because our rockets travel as slow as they do, and the Mars advocates want to do something NOW – within their lifetime. Alternatively, if Mars was truly suitable for colonizing, there wouldn't be an interest in exoplanets or speed of light travel. Mars is a compromise, a desperate compromise, influenced more by emotions, jobs, and more than a touch of a messiah complex, than any purely scientific rationale.

This partly answers the question "Why Mars?" Outer space exploration is limited to our own solar system due to distance issues, and even within our own solar system, distances are vast, but Mars is the next planet over. So the Mars advocates have presented the argument that it's simply the least inconvenient of all the planets to get to. The... *least inconvenient*. This may justify getting an unmanned rocket there within the limits of our patience – but this is not sufficient grounds upon which to base a permanent, human colony, given all the other completely unearthly circumstances. Not even to send a party on a round-trip scouting mission. No invincible arguments exist to support this – none whatsoever.

A rocket that could get to Mars in a few weeks, which could be launched any day of the year and not once every 26 months, would be necessary to make even a purely scientific camp worthwhile in the same sense that scientific camps are worthwhile in Antarctica. In which case one might imagine a colony of a few hundred on Mars, more with round trips. Such a rocket would be the fourth biggest technical breakthrough of all time, after the car, television, and nuclear energy, where going to Mars might become a real prerogative for the average scientist (or even space tourist!) – or lead to horrors greater than ever imagined – like at the top of Everest, bodies frozen, eyes staring. And it should be understood that today's rockets are the same as they've always been, relying on the same old "rocket engine," in the same way that cars haven't really changed from their first inception, relying on the same old "car engine." And in the same way a car can't be designed to "just go" 10 times faster, a rocket can't "just go" 10 times faster – and rockets would have to go about 1,500 times faster to travel the distances we're talking about, at 15% the speed of light (using 70,700 mph as the contemporary reference point, the cruising speed of the Cassini probe). An entirely different engine would need to be invented – such an engine would be fantastical and couldn't be "just invented." There are several types of "faster" engines in the minds of engineers, some exist as models or prototypes, still others are on the drawing boards. But the rockets of tomorrow will not simply be a matter of engineering chutzpah. They will require scientific breakthroughs that are as yet unimagined, rely on completely different schools of thought, a completely different perspective of space and time, and perhaps even the reinvention of science itself – or may be impossible.

The study of exoplanets further than those already discovered seems to be a complete waste of time, because even if an exact duplicate of Earth – or something even better – was discovered, say 50 light-years away, the problem is that's 50 light-years away! We would only drive ourselves crazy, especially if an apocalypse were to happen, trapped here and denying the inevitable. The reason we're focused on Mars is because it's within the realm of possibility that we can even get to it, not because it might be habitable, the truth of which is becoming more evident with each successive unmanned mission, which should be obvious simply because each successive mission continues to be unmanned, but sometimes it's hard for those with an ideology based on hope to see the obvious. This is not to say that unmanned research is a big waste – on the contrary.

THE FACT THAT THERE ARE BILLIONS OF EXOPLANETS IN OUR GALaxy doesn't increase the odds that we would ever find or visit one that is Earth-like, it's only of interest to astronomers because it represents a major jobs program for them – it gives them lifetimes of busywork to do, so aren't they glad! But beware their self-serving press releases, because of these billions only a few could ever be relevant to us – those belonging to the closest stars – and they're all known to be not habitable. So that's the end of that! It wouldn't really matter if exoplanets further out are habitable because they'd take too long to get to. It would take a whopping 70 years to get to an exoplanet as relatively close as 10 light-years away travelling at 15% the speed of light, which is 100 million miles an hour, or about 1,500X the speed of today's slow poke rockets – but travelling at 15% the speed of light is, and will probably always be, science-fiction. This equates to about 95,000 years using today's rockets, travelling at 70,000 mph. Solar systems are tremendously distant from each other, they are light-years apart, so you see how impractical it becomes just to count, much less investigate, exoplanets further than the dozen or so closest to us – and we already know that the ones further out than the closest ones are not habitable. And it's not as if a new batch will cycle through every 20 or 100 years, so the lack of prospects regarding this, as an alternative to Mars or something better, is, like it or not, final.

We'd have to study perhaps thousands of exoplanets before finding one suitable to colonize, send unmanned probes, etc., which would take centuries of travel time (even at 100 million miles an hour) per suspected Earth equivalent just to get a single high-speed probe there (compared to the 8 months to get a probe to Mars). Which means it would take the same number of decades or centuries to send *manned* spaceships, probably longer – travelling at 100 million miles per hour – but nothing we have can travel that fast. In contrast, if a ship was travelling to Gliese 581d (the closest, most "Earth-like" exoplanet) at the more realistic but still disappointingly slow 70,700 mph, it would take almost 190,000 years to get there. That's the cruising speed of the Cassini spacecraft launched in 1997 to visit the outer planets of our solar system. So forget about that. But it's Earth-like! FORGET IT! This of course makes the 8-month trip to Mars seem easy – barring breakthroughs in spaceship design, which aren't really even on the horizon – despite the brainstorming, which should tell you something.

BUT REST ASSURED. IT'S REASONABLE TO THINK WE'RE ALREADY on the best planet of them all – Earth. This may not be an accident or the result of some frivolous ratio of chaos and order, but an intentional, best possible choice of an intelligence greater than ours. Either way, we're arguably in the best possible place to be, or "the best of all possible worlds." Some will object to this, pointing to the abundance of evil or adversity that is impossible to reconcile – but it seems there are some things we can not run from, even if we could get to another planet. Both Thomas Aquinas and Voltaire, and also Leibniz, pursued the "best of all possible worlds" argument from a different perspective, regarding good and evil, but this provokes the question of whether or not we should pursue exoplanets that are merely livable (or not livable at all like Mars), or ones that are even better suited to human existence than Earth; we certainly don't want to fling ourselves to hell as a deliberate choice.

Do you see how I'm considering every possible angle? I've summoned the ghosts of Leibniz and Aquinas and there's still nothing – Nothing! – on the plus side for Mars.

There are literally too many exoplanets for us to ever know for sure whether or not there is an Earth equivalent out there, or how many. If we could send out probes – at the speed of light! – to other solar systems, or even galaxies, we could never investigate them all – maybe not a single one. Crazy Far – that's the title of the National Geographic article (January 2013) mentioned above, with the subtitle "Will we ever get crazy enough to go?" Maybe the crazy-tide is about to turn. Using space-based telescopes allows us to learn, but has really only given us insight into how vast space really is.

Manned space flight is a complete waste of time.

At this point in our conquest of space, our preoccupation with manned space flight has become outdated and our quest for knowledge would be further along if we devoted ourselves instead to *unmanned* exploration of the stars. Which leads me to think that manned space flight is a complete waste of time. Advances in miniaturization, robotics, computerization, and artificial intelligence make such a paradigm shift seem obvious. One might, then, consider this – unmanned – exploration to be the pinnacle.

There's a reason the various solar systems are so far apart – their longevity depends on their isolation, are a function of it. The solar systems and planets within them, and the galaxies they comprise, need to be so far apart that they would never crash into each other, since they're already moving, orbits wobbling – even on this scale there seems to be a self-preservation – a higher intelligence – at work. Even when galaxies seem to collide, it's more accurate to say they pass through each other. We should be grateful we are light-years from the closest exoplanets, rather than defy the odds in getting there and beyond. In defying the physics that keeps us here, it may be impossible for us not to destroy ourselves – what grand creatures we must be to think we should.

It may be a greater freedom to simply embrace our mortality, and the relative tranquility of what we already know.

48

Goldilocks Zone

No human born on Earth will ever travel to a planet in another solar system.

Nor will we ever have sufficient knowledge of some distant exoplanet to give us the certainty that such a one-way trip would be worthwhile. And no one would ever set of on a journey lasting a lifetime, or generations, without absolute certainty.

The time frames involved in exploring the closest exoplanet using today's rockets for unmanned reconnaissance – would span one thousand human generations.

No two Earth-like exoplanets are so close that any human will set foot on both of them.

Earth-like exoplanets (actually supporting an Earth-like ecosystem) are at least rare enough that to get from one to another will always exceed a human lifespan. This means no human will ever know more than one Earth-like exoplanet.

To spare further torment, these statements should be treated as axiomatic.

I'M JUST PULLING THESE TRUISMS OFF THE TOP OF MY HEAD, THEY require absolutely no deep thought or calculation – but which you will never hear from an astronomer or anyone in the space community, they're just glad to have a job. Here are some more:

No human will ever travel at – or near – the speed of light.

Even unmanned craft will (probably) never travel at – or near – the speed of light – which would be required for any craft to reach even the closest solar system.

Humans will never travel distances measured in light-years.

Any craft launched toward another solar system – and which would be travelling at much less than the speed of light – would be forgotten long before it got even halfway there! This includes the Voyager 1 probe, launched in 1977 towards the edge of our solar system and beyond, which would not reach even the closest solar system, Alpha Centauri, for 78,000 years (if it were headed in that direction). The Voyager 1 actually did leave our solar system and enter interstellar space in 2012 (this is the probe carrying the gold record inspired by Carl Sagan), which means that after soaring through space at over 35,000 miles/hour for 35 years *it was still inside our solar system*! – which surely indicates the real limitations to learning about the planets and other things just within our own solar system.

Forget manned or even unmanned craft. Electronic signals that might be sent to planets in other solar systems – exoplanets – would take years to get there – centuries. The distances involved are literally beyond human comprehension. One must ask, even if I could travel such distances, why would I? Especially if it means giving up regular food and sex – you know, being human. To make such a sacrifice there would obviously have to be tremendous practical benefits. On a somewhat smaller scale, it's like climbing Everest. I could, but do I? And once there, do I remain? Why do we not establish a colony on top of Everest?

Imagine your ship is travelling through deep space and a voice comes over the loud speaker. "We'll be entering the Goldilocks Zone in 3 hours." Will that mean hospitable (and that I can wear my khakis), or just habitable (and that I should wear my spacesuit – in which case I hope it's sexy because when I'm arriving on a new planet I want to look my best).

Earth 2.0 (a hip name for an imaginary Earth-like planet – such as one might find in a so-called Goldilocks zone) is really quite a contrivance. By implication, Earth 1.0 would be good old planet Earth, of course, but why belabor the issue? I don't want to be a scold, but listen people, THIS IS IT! Stop looking for a Goldilocks planet – WE'RE ON IT! And we must make the best of it. I mean, what do you think, that the Earth has a twin out there, in some neighboring solar system, that we'll all just conveniently shuttle ourselves to when we've burned up all the gasoline here or discovered there's nowhere else to put all the indisposable waste we're creating? Even if distance were no object, a planet that's 90% like ours would still be basically unlivable. Or so inconvenient, our physiologies unable to adapt, we would go crazy and commit suicide. As I suggest elsewhere, even the most apocalyptic vision of Earth is at least ten times better than what's *out there*.

It's amazing how many people think it's inevitable that the most bizarre science-fiction should become fact, which seems to be the premise for the Earth 2.0 concept.

One of the weird things about Goldilocks planet hunting is that it would need to be Earth-like *now* to be a Goldilocks planet. Earth wasn't always the way it is today – habitable – and this is true for other planets that may be habitable. Many planets out there may have been habitable in the distant past, or might be in the distant future – but we're not interested in those, and that narrows the odds of finding a true Goldilocks planet, or Earth 2.0. The odds of finding an Earth-like planet now, i.e., a planet that is, in its present state, Earth-like, rather than a planet that might be Earth-like in a million years – or had been but now isn't – is similar to the twisted logic of how likely we are to be visited by aliens *now*. You could say that the odds of the next exoplanet being Earth-like are similar to the odds that the next flashing light we see in the sky is aliens bringing us gifts. Given the overwhelming enormity of space and time, the odds of either are basically zero – and I'm emphasizing the "in our own time" part, which you may construe as your lifetime or the duration of the human species, either one, it hardly matters. And it's hard to trust astronomers who say that an exoplanet is Earth-like when they claim Mars isn't just Earth-like but "hospitable!"

We may discover that planets are like people, where no two are alike. We are each one of a kind, the same is certainly true for planets – making Earth, of course, truly precious. We've been studying Mars for years, up close and personal, and we've suspected that Mars is basically uninhabitable. This was true even before the four rovers, which have simply confirmed what theorists had assumed, or that we could deduce from satellite photography alone, so it seems ridiculous that some scientists are still unsure. Questions get answered, but we keep asking the same questions, *as if things might change overnight.* As if the average temperature on Mars would be different today than it was three years ago. As if the day time temperature at a certain location on Mars was -200 F last year but 70 F now. It seems that at least one big question has been answered: there are no Martians on Mars.

It's almost as if the planets and solar systems are far enough away, in the context of a human life span – which is a fundamental point of reference – such that a mass exodus to another planet would be impossible *by design*, because a so-called Earth 2.0 would have to either be in our own solar system or the closest one to us, and we know this is not the case. It bears repeating: there is no limit to which the habitability of Earth could degrade (bearing in mind that natural circumstances seem to favor a certain *hospitability*) that it still wouldn't be infinitely better than any Earth alternative – or Mars – if the long-term survival of our species is the goal. Moreso if you expect the quality of life, comforts, and modern conveniences that we take for granted – and which make us human.

No human born on earth will ever travel to a planet in another solar system. Nor will we ever have sufficient knowledge of some distant exoplanet to give us the certainty that such a one-way trip would be worthwhile – *and this is certainly no less true for Mars.* And no one would ever set of on a journey lasting a lifetime, or generations, without absolute certainty. And the further the destination, the more certainty you would need. But it's funny that, since all exoplanets are equally unattainable, I would suggest that some are close and some are far! "Close exoplanet" is an oxymoron. Even travelling at one-tenth the speed of light it would take 42 years to get just to the nearest solar system but we already know that doesn't have, in any sense of the word, an Earth-like planet. End of discussion. But even if it did (although to determine this with certainty would require technology that does not exist, much more than the practically useless classification as a Goldilocks planet if such were the case), and we decided to colonize it, two other key requirements must be satisfied: the astronauts would need to be in a state of suspended animation, and the rocket would need to be completely controlled by an artificial intelligence far superior to anything that has ever been imagined (both of which are currently impossible and in the same arena of bizarre fantasy as terraforming) – but no one would give up 42 years of their life for this – they would not even live long enough to survive the journey.

Suspended animation would be untestable and is probably more akin to bringing back the dead than we'd like to think, and any artificial intelligence smart enough to provide a fail-safe means to some distant Earth 2.0 would devise more practical strategies for our long-term survival on Earth than to leave it.

To first determine unequivocally that a target Goldilocks planet was truly Earth-like would require that reconnaissance satellites be put in orbit around it and possibly even rovers like on Mars, but the time frames for these – to explore just one of the closest exoplanets using today's rockets – would span *one thousand human generations* – and I don't mean in some poetic sense that it can't be done, I mean 40,000 years literally – which equates to unfeasible – and during which time mankind's priorities *will* change – our priorities have certainly changed since 38,000 B.C. – so scrutinizing Goldilocks planets even further out would be exponentially more unfeasible. Jesus Christ was born 2,000 years ago, but you want to send a rocket on a trip that is twenty times that? Even if in the meantime we built a rocket that was just twice as fast it would get there long before the first one, so you see how ridiculous things get when we start sending things to exoplanets. There would be no point in sending even one probe to the nearest exoplanet until we could build the fastest possible rocket that is humanly possible. This would require the postulation of some theoretical limit – but which may be impossible if we're relying on hubris ridden rocket scientists.

IF WE WERE LOOKING AT OUR OWN SOLAR SYSTEM FROM JUST 20 light-years away, astronomers would probably classify Earth as being too close to the sun – and uninhabitable – and Mars as a Goldilocks planet! – and never know the difference! This demonstrates the absurdity of the whole thing, or at least an *uncertainty factor* that may be unavoidable over distances measured in light-years. Because except for a few million miles – which is nothing even within our own solar system in terms of a planets distance from its sun – Mars *might* be livable, if not for humans or an advanced ecosystem – something! It's only because we're actually *here* and not 20 light-years away that we know the difference. So there's a limit to how much we can trust ourselves to analyze data from light-years away. Instead of Goldilocks *planets*, we might better be looking for Goldilocks *solar systems*, in which there would be *several* potential Goldilocks planets – which would double or triple the odds that we're at least looking in the right direction – odds that might make the whole thing seem a little less silly.

Even rovers like the ones we've sent to Mars would be somewhat untenable for exploring exoplanets. Ground control operates the Martian rovers via remote control, and it takes an average of about 30 minutes for each set of instructions to travel from Earth to Mars, so the instruction "move forward 10 feet" sent to the rover isn't received by the rover for 30 minutes. But for an exoplanet 4.2 light-years away (the distance to the closest exoplanet), this instruction wouldn't be received for 4.2 years. So for planets outside our solar system, by which I mean ALL exoplanets, remote controlled rovers would be useless. Even if a rover was completely automated and didn't rely on instructions from Earth, and we had satellite reconnaissance to help us design it (as with the first Martian rover), the degree of artificial intelligence this would require would be astounding. A scenario involving such a superintelligent rover on the closest exoplanet is only as likely as very high speed rockets, like at least 15% the speed of light, or 100 million miles an hour, which would take 28 years to travel. Would anyone have the patience for this *already knowing the closest exoplanet isn't Earth-like?* Even with such a rover on the nearest exoplanet automatically sending data to Earth (at the speed of light), the time frames involved in receiving this data would be mighty inconvenient. We wouldn't even know until 4.2 years after it landed...whether or not it landed! In terms of relative feasibility, getting a man to Mars would be easy compared to getting a rover to even the nearest exoplanet. So knowing that it's currently impossible to get a man to Mars, it's easier to appreciate the unlikelihood of sending a rover to even the closest exoplanet.

> **Speaking of bringing back the dead, it would seem this and Earth 2.0 have nothing in common, but both would seem to provide the means, if in radically different ways, to achieve what the Mars advocates want: immortality.**

What would it take to get the Mars advocate's priorities, if not all of mankind's, to turn in this direction? We sure wouldn't be confined to some damned four-week launch window every two years, or unending dial-a-flavor. Just don't say zombie.

The only possible scenario involving humans living on an Earth-like planet in another solar system – but which can only be imagined – would be purely altruistic, and involve somehow encapsulating microscopic pieces of human life, such as in the form of DNA – and in a manner as yet unimagined transport this to the Earth-like planet in that solar system. One might imagine millions of such seeds, like a stream of photons. But the success of this or any other plan would involve an enormous amount of luck *and we would never know if the plan worked* – and it probably wouldn't.

But altruism could never be held responsible to fuel such a fantasy. It is also true that capitalism does not reward altruism. Given this reality, the prospect of humans on another Earth-like planet is and will remain impossible for a long, long time.

49

Aliens

Skepticism doesn't end when I confront the alien.
Skepticism begins when I confront the alien. Anything before that is inexcusable nonsense.

Do you think "aliens" wouldn't have the same logistical
issues in travelling through space – regarding distance
and time – as we do?

WE TAKE FOR GRANTED HOW INTELLIGENT A SPECIES NEEDS TO BE
in order for it to develop space travel. Even on Earth, and even for the
millions of years that Earth has been crawling with countless species
(with intervening periods when Earth was hostile to life) it's only in
the last fifty – yes, that's fifty – that there has been a species intelligent
enough to develop space travel! Except for that blink of an eye, in the
cosmic sense, that we're living in now, the odds that such a thing would
happen are practically zero! But no matter how slim the odds are that
some distant planet may harbor even the simplest forms of life – that
may even be *crawling* with life! – the odds are many times less than
that – practically zero – that any species would be intelligent enough
to...develop space travel! Odds on the order of 1,000...(many zeros) to 1.
After all, how many species were on this planet before one came along
and developed space flight? Millions! So it makes sense that only on a
planet that already has millions of living species – a hierarchy like the
animal kingdom we have on Earth – could there be a *single* chance that
there *might* be one advanced enough to develop space travel. These are

enormous leaps, going from "harboring life" to "harboring intelligent life" to "harboring life intelligent enough to – of all things – develop space travel." And then of all the infinite directions, to head in *our* direction, *towards Earth*, at *any* time during the course of our existence? The odds of this final leap happening are, of course, basically zero; whether that's a "probably zero" or a "definitely zero" is more a matter of intuition than calculation. Realize of course there is the additional matter of proximity, because you can't ignore the time it would take to get from one star system to another, no matter how fast one is travelling. This is what we're talking about in terms of the likelihood of being visited by aliens.

> **It seems one might even question the unquestionable assumption that an intelligence capable of developing space flight would...develop space flight!**

Or be as obsessed with technology as we are. And the odds that some distant planet may harbor life of any kind whatsoever may be even *less* than I'm suggesting – such that the odds of any interaction between us and an alien race are simply not worthy of the concern we've invested in it.

Even if we assume that intelligent beings exist right now on another planet, and that they're aware of Earth and are intrigued by it, and have space flight technology that exceeds our own: sightings of UFO's or mysterious lights in the sky, darting around like photons, would more likely be probes or unmanned drones from other planets than actual aliens, in the same way that we send unmanned craft to Mars, or all the other unmanned things we launch into space – are more likely this than the larger craft that would be expected to carry a crew, living quarters, and supplies. And in the same way that we don't yet have the technology to land heavy craft gently on Mars – which would be needed for humans to land on Mars without being killed by the impact – curious and intelligent creatures on other planets, even if they're transcendent to us in some ways, would most likely be limited in other ways, such as...in their abilities to send *manned* craft here – as if the manned/unmanned dichotomy exists for alien intelligences as for us – which may be a bigger assumption than we realize. But assuming this dichotomy is in fact universal, it's one thing to send unmanned spacecraft from one planet to another – even we can do that! – but manned missions are a completely different story. Just *landing* is – obviously – a completely different story. Who would have thought one could get all the way to another planet...and not be able to land! Certainly for us and Mars, it's *landing* the damned thing that's the problem! It's those last few inches! So close and yet so far. But why do we think that beings on other planets wouldn't face similar challenges, or similar barriers to living on another planet? They may be even more fragile than we, or have metabolisms finely tuned to their home planet with no tolerance for variation...or have allegiances that forbid them from leaving their planet, or whatever other contrived or natural limitations or *hang-ups*. Maybe we're in *the forbidden zone* – who knows? Maybe they're just more careful, or more timid, just not as adventurous as our "Right Stuff." Maybe our alien counterparts don't have "eye of the tiger" or that "killer instinct," the "need for speed."

> **A testosterone fueled sense of conquest or promiscuous need to break barriers is very likely a uniquely human quality – rare even among humans! But this is just a single dot on the spectrum of the human condition – an incredibly unlikely dot – a dot that is certainly unique to Earth, to humans – we would surely never find another example of it anywhere.**

Which is all another way of saying that intelligent life on other planets doesn't need to be analogous to intelligent life in Earth – and probably isn't. What's interesting about this manned/unmanned dichotomy is that we continue to put manned exploration above unmanned, a priority that might be completely reversed in higher intelligences after all.

Yes, it's conceivable – or easy enough to imagine – that some distant, alien *entrepreneur* might send a tiny, unmanned, speed of light probe this way to collect and return data. Unlikely but conceivable. This doesn't contradict what I said in the previous paragraph, where I simply argued against aliens sending *manned* craft – and would answer a lot of questions.

But one thing is certain: it's completely unrealistic to think a being from another planet could just travel some huge distance, then just land, then just walk out and say hi.

This could only happen in the movies, and in our imaginations. Some of them, at least, regarding the UFO sightings again, are most likely uncommon but natural (some would say spooky) phenomenon given meaning – supernatural meaning – by the human tendency to give things meaning, and inflated by our imaginations. And accepted as truth by unscientific or simple minds. So imagine what happens when you add to that a dash of paranoia or wishful thinking. We used to think there were little green men on Mars. The reality is, of course, disappointing.

Either way, the general wisdom regarding aliens visiting or living on Earth, which dates back many centuries, is due for an overhaul. Three distinct themes have dominated our myth and culture, and which are embraced by Hollywood. Aliens would either one: enslave or otherwise subjugate us, two: annihilate us using a sophisticated but brutal war-craft not unlike our own as a prelude to full ownership or three: exist in a harmonious relationship with us and all of Earth's species in a way that might be too subtle to ever be noticed. And if they *don't* come it's because we're "not ready" (how kind of them!). These scenarios persist because they capture the imagination and are bankable. But none is likely. It's far more likely that they would be indifferent to us, or afraid of us, the same way we're afraid of almost all creatures on this planet even though they all rank far below us in terms of intelligence. In the same way, any alien that may get this far would be as afraid of us as we are of hornets and spiders.

One way to ascertain the likelihood of being visited by aliens – assuming there are intelligent beings on other worlds that have rockets – which is of course very un-likely – is to ask yourself, would you travel to theirs?

We have rockets – you want to go? Aside from some very serious practical matters, what it boils down to is, how much time would you be willing to give up from your short life to travel to another planet? It seems anyone would actually be *less* likely to travel to any planet believed to be home to a complex ecosystem of intelligent creatures, but this would be impossible to know in advance, beyond the basic question of habitability. Because, being light-years away, any degree of reconnaissance would be impossible. Time is the critical factor, not distance, because, as man has shown, we can make vehicles go faster and faster – which effectively shrinks distance, but we can't just add years to our life – a year is a year and once spent is gone forever and we only live once. So how much time are you willing to give up? You probably wouldn't be willing to spend 50 years on a rocket travelling to another planet, even if unconscious.

> **You probably wouldn't be willing to get into a rocket at age 18 with the expectation of arriving at another planet at age 68, there is just no acceptable combination of patience, priorities, and profit that would make this a satisfying or worthwhile experience. And this would probably be as true for our astronaut counterparts on some distant, backwater exoplanet, gazing up at the stars as they ponder the infinitude – while at the same time embracing their comfortable indifference.**

This is one of the minor reasons why we could never be visited by creatures from other solar systems:

> **The more likely an alien species is to have the technology to come here (and therefore the more reasonable they would be), the LESS likely they would be to actually make the trip – and which seems to be axiomatic.**

> **Any creature that could travel to Earth from another solar system – moreso from another galaxy – would be transcendent – and most likely indifferent to us, and in the most extreme way possible!**

It'd be like you walking to the southern tip of South America to visit a termite colony. In other words, ridiculous, regardless of any capability to do so.

> **There's just no reasonable need, or rationale, for a species as transcendent as one would be that could travel this far, to travel this far.**

And how vain, or paranoid, must one be to think that such a creature would travel this far just to contact *them*, as some think.

Such a trip would take an indefensible amount of time and energy, or some loathsome compromise between the two, such that it would always remain somewhere between pointless and impossible. Science-fiction can ignore these realities – and always does! – but we can't, nor can transcendent creatures on other planets blissfully ignoring us.

THE COSMIC LAW OF INDIFFERENCE

In addition to all this, and in formal terms, there may be a *fundamental inability for creatures from one planet to live on another, insofar as the constituent parts of a living thing are – more than we might ever know – a function of the planet that gave rise to it.* I won't give an official name to this theory, but a corollary to this might be called the *Cosmic Law of Indifference*, in which there shall be no interaction between living things from different planets that would be mutually beneficial – as if one might dovetail into the others ecosystem or life-cycle – but instead, to the extent that there is an interaction, there would be mutual disruption and destruction; it would simply be a matter of degree. This could never be proven scientifically (as the scientific method is currently defined), however it seems the case can be made, and may be one reason why there is so little evidence, that is, no evidence, of alien visitations, notwithstanding so-called UFO sightings.

A lack of evidence considering the preeminence and prevalence of our species.

Any creature smart enough to get here might be alert to this fundamental problem, or have discovered it – which is not to say that aliens have ever visited Earth, given the infinitude of space and time. This fundamental problem wouldn't affect our ability to travel to and visit nearby planets, but it would certainly curtail our ability to live on Mars, and the prospects for a colony on Mars, despite our greatest efforts.

MCKINNEY'S FIRST RULE OF INTERPLANETARY TRAVEL

Another, simpler way to describe the Cosmic Law of Indifference is through *McKinney's First Rule of Interplanetary Travel* (and which may be considered tautological to the Cosmic Law of Indifference): You can't move to another planet. This means you can only live on the planet you're born on. It's the old "fish out of water" routine. Which may be another reason the aliens that supposedly fly by Earth don't stop – they've learned this, and know enough not to test this galactic law even just to stay for a quick hello. What's born on Earth, stays on Earth. Whether known or unknown, there are just too many adversities to manage already.

This means that nobody who goes to Mars will survive there. The issue ultimately isn't one of technology or science, but human nature.

Many will object to my Cosmic Law of Indifference – such as those with abduction stories – but others too – as if aliens are a proxy for God or a big brother or father figure that's missing in their lives. Certainly the aliens are not *indifferent*! Many people need to think that "someone" is watching them - *watching over them* – that this someone is benign and will ultimately, eventually, connect with and possibly even save them. The "a transcendent being needs to contact me" syndrome.

I wonder if any of these people are as tormented by the "I need to walk to the southern tip of South America" syndrome as the aliens are tormented by the "I need to contact a human" syndrome.

We've all heard the "we were out drinking" stories. Leave out the obligatory "we were out drinking" part and you'd have a much better story. And this is what happens with most if not all abduction and UFO stories – they leave out the "we were out drinking" part! And would you believe them otherwise? These people are just looking for attention (which may in turn be related to more deep-seated psychological problems). For some, being even just a little "under the influence" is so normal that it wouldn't occur to them that their abduction or UFO experiences are imagined or dreamed – and of course inspired by all the other identical stories that begin and end the same way.

IF SOME FAR AWAY CREATURE WERE LOOKING FOR ANOTHER PLANet to live on, and visited Earth – do you really think it would want to stay? Earth couldn't possibly be their first choice. There are really quite a few things going against Earth for a species that would clearly have the means to travel among the stars if they got this far (because they obviously didn't come from within our own solar system). It's like when you're buying a new home, do you want something old, or something new? It's the same with planet shopping. Any advanced species (with the means to do so, which is astronomically unlikely but regardless) would select a planet that's at the beginning of its life, not one that's as middle-aged as ours – or on the verge of apocalypse – and overrun with all manner and variety of slimy or hairy or trigger happy species which for the most part don't hesitate to devour each other and in many cases live inside each other! How horrible is that? As if Earth was nothing but a giant disgusting incubator of life! And all the areas any discriminating alien would want are already in use. In other words, they'd reject Earth because it's used. Earth is used up and overcrowded. Why would an intelligent species – one with a planet shopping list in hand – want to start out anew with that? In real estate terms (and is there any other way to look at it?) we're not "on the market" – and we need to build taller buildings and transmit more noise into the void to make this more clear. You wouldn't buy a house with a family living in it with the intent of kicking them out, would you? Of course not. It's the same with planet shopping. It's not as if they'd be desperate – with their home planet breathing its last gasp, or about to be consumed by its own sun! Or another one of those "just in the nick of time" scenarios. Any intelligent alien species (with the unlikely means to do so) would be shopping around well in advance of its own imagined extinction – proactively. We would simply be one of many on their shopping list – and would more than likely be snubbed than anything else. They might "taste test" us and spit us out in disgust – this may have already happened. Maybe they've landed and stayed for a while, built some pyramids, carved some crop circles, or whatever, and left. What better reason to travel among the stars than to build pyramids and then split? As if highly evolved creatures, including ourselves, share a universal discontent with being in one place.

One might think that if they've come this far they would consider Earth to be the grand prize, with its raging excess of life forms, and gorge on the feast – letting nothing stand in their way – in terms of sampling anything and everything from the animal and plant kingdoms. But there are three reasons this wouldn't happen.

ONE: spacecraft coming from other star systems would probably be thin and light, with no room to spare to carry back much, if any, in the line of souvenirs.

TWO: anything that's alive would probably die in the time it took to get back to the aliens' home planet – not survive the rigors of interstellar space travel.

THREE: since there shall be no interaction between living things from different planets that would be mutually beneficial, as I proposed earlier, and any transcendent being would know this, they would be abhorrent just to the idea of taking any living thing back with them. And things brought back to whatever aliens' home planet, especially living things, would probably be as upsetting to its residents as we are disturbed by the thought of extraterrestrial life creeping around on Earth!

THE ROUND-TRIP FALLACY

But if they've come this far, and have settled in...they probably wouldn't leave. Due to the "round-trip" fallacy. What's interesting about the alien visitation scenarios is that they all include a round-trip for the aliens! If they were able to adapt to Earth, and live among Egyptians, and build pyramids – or not...either way it wouldn't make sense for them to leave (the time and distance issue again, no matter how fast they could travel), they would just stay! (Which of course suggests that if aliens are not here now, they've probably never been.) Like the plan for going to Mars, which some unequivocally state would be one-way.

> **With interstellar travel (as opposed to the more short-range interplanetary travel), it seems that all the time lost to always being in between places would not be compensated if one just turned around and left.**

Do we think our interstellar travelers have nothing but time on their hands? If you can't remain at the place you've worked so hard to get to – which would be the case for aliens who might come this far – from another star system – why bother going in the first place? It just isn't reasonable, and any creature of interstellar space travel would be *in-*

credibly reasonable – which means they would stay. It may not even be a choice! Rather, it seems that a basic tenet of interstellar space travel is that it be one-way, due to time constraints that would exist even travelling at the speed of light – which should be taken to mean some enormously high speed, but considerably less than the speed of light, because speed of light travel is a goal that man (or any intelligent creature) will never realize, due to natural limits and not lack of creativity. On the other hand, travelling within a solar system wouldn't necessarily be one-way, but probably, for the same reason, even if one could travel at a "enormously high speed."[1] Alien space travelers wouldn't be relying on the promise of a round-trip, i.e., a return to their home planet. That is, after all, the idea when creatures travel huge distances: to move. That it be permanent. One-way. Isn't there also a natural sense of repudiation toward the place one is moving from? The point may not be so much to get to Earth (or wherever) as to get *away* from their home planet! – which is the case with the Mars advocates and Mars, to a large degree. It's a mental hang-up to think that everything must return to some previous state, that the status quo can never change, that any alien that might visit Earth must necessarily return to its home planet.

> **The rule when travelling – or moving – to another planet seems very likely to be "adapt or die." This would be as true for humans on Mars as aliens on Earth.**

The Round-Trip Fallacy is especially relevant to time travel or speed of light travel, where the idea of returning home (to your wife and kids) wouldn't just be unlikely, but have no meaning whatsoever. Which of course should be a strong deterrent to anyone wishing for either of these.

To understand the Round-Trip Fallacy requires an appreciation of the element of time, but time is usually ignored in science-fiction novels and movies and so how are we ever going to learn about it? So the Round-Trip Fallacy lives on, where the characters always return home and kiss their wives and children after some great adventure involving

---

1   Actually, for travelling within a solar system, the maximum desirable cruising speed would be much less than the speed of light, considering the time it takes to accelerate and decelerate, which is greatly limited by our ability, and the cargo's, to tolerate the G-forces that would be experienced during acceleration and deceleration of a spaceship.

time-travel or travel to some distant planet, which would not be possible in real life (moreso for interstellar travel). And so the Mars advocates assume this will be the case with Mars – or eventually be the case – getting back to Earth. Because in real life, people want to "go home." It's part of the human condition. Even E.T. at least wanted to phone home. The lunar astronauts never had this problem – never encountered the Round-Trip Fallacy – it just wasn't an issue for something that close to Earth. But it's definitely an issue with Mars and anything further.

ALL OF WHICH DEMONSTRATES – LET'S CUT TROUGH THE PRE-tense – just how stupid astronomers and so-called scientists can be: Some have not only proposed, but insist upon, the idea that the pyramids were built by wandering bands of supremely intelligent aliens. There are literally dozens of books on the subject. Why is it that because today's engineers can't figure out how or why the pyramids were built that they must therefore be the work of a superior intelligence that takes the form of planet hopping nomads or wandering bands *from another planet*? Which begs the question, why do we think that extreme intelligence from other planets would take the form of *wandering bands* when there is nothing about intelligent creatures on our own planet that would suggest this or from which one could extrapolate such behavior – including us? One could only conclude the opposite, from what we know about ourselves – that superior intelligence leads to selfishness and materialism. So how could it be the nature of otherworldly extreme intelligence that it would allow itself to be *consumed* by travel time, possibly in some form of suspended animation – which certainly seems to be a colossal waste of time – but which would have to be the case given Earth's distance from other solar systems – all for the sake of building something as useless and meaningless as a giant pyramid! When our astronauts went to the Moon, did they build anything that would be equivalent to an Egyptian pyramid? The answer is an embarrassing "no." The litmus test of superior intelligence: join a wandering band of comatose space travelers to build giant but meaningless geometric objects on other planets. Who would not envy such a leisurely existence? It might be one of mankind's most embarrassing shortcomings that we haven't yet achieved this degree of leisure.

What's interesting is, why would wandering bands of aliens know anything more about building giant pyramids than ancient Egyptians? Even though the consensus seems to be that Egyptians didn't have the know how to build giant pyramids, I'd like to conjecture that...neither would aliens.

Some have proposed that aliens may have built the pyramids as a kind of landmark. So that if and when they ever returned, on subsequent visits, they could more easily return to that spot – as an aid in *celestial navigation*, as it were. But it seems that would be a lot of work for nothing – worse: trading a small inconvenience for a much larger one. Because any intelligent being that could navigate to Earth from another star system would have already mapped out the Earth and its countless "points of interest," which would make returning to any specific location on Earth easy enough, including the location where the pyramids stand today – so they wouldn't need to actually (of all things) build something as enormous as an Egyptian pyramid – or several in the same location! It's not reasonable to think that planet hopping nomads would need an Egyptian pyramid in their celestial navigation, since the Earth is constantly moving and would not be as good a reference point as, for example, our sun, which, due also to its brightness, even to whatever visiting aliens' home planet, could guide the aliens to Earth. Navigation savvy aliens (and is there another kind?) would use primarily stars to navigate, not planets, certainly not something *on* a planet. Only the location of Earth, but nothing on the Earth, would affect the flight path of a starship headed our way. So in returning to Earth they would first zero in on our sun, and then our planet. This would be their strategy, which is complicated enough. Once they reached Earth, getting to the area that might otherwise be marked by pyramids would not be facilitated by the existence of pyramids, they would need maps anyway. At the point they'd be visually scanning the landscape for pyramids they'd already be at the general target location – it's not as if there was some critical spot – upon which must be built a giant pyramid! – that would have interested them. If any spot on Earth was that important to them they would have never left in the first place and just stayed – which of course also implicates the Round-Trip Fallacy again. One or a dozen enormous pyramids sitting stupidly in the desert would not be a reasonable substitute for any visiting alien's navigation and mapping systems. So the idea that the pyramids were built as a landmark by aliens to make a return visit easier is outrageously ridiculous. Otherwise there might be pyramids throughout the galaxy, like a trail of breadcrumbs left by wandering planet hoppers. Would you like to think that?

> **One or a dozen enormous pyramids sitting stupidly in the desert would not be a reasonable substitute for any visiting alien's navigation and mapping systems.**

Here's the answer to the mystery of the pyramids, regarding the phenomenal task of their building: the ancient Egyptians were smarter and more resourceful than we think, or, at least one of them was! It seems one is forced, given all that we know, to use this Occam's razor logic. Every era has its Einstein. We have ours, they had theirs – if perhaps not actually an Einstein. Even if the answer is more complicated, belonging to the supernatural or paranormal, intervention from planet hopping nomads should be considered outside the realm of possibility for the reasons stated concerning the infinitude of time and space.

MISSING NUTRIENT THEORY

There could, arguably, be a reason an alien would come to Earth: our planet has a resource it needs. It had used up a nutrient on its home planet. But at the same time it seems this would never happen. We will never have to travel to distant planets in search of critical stuff we need to live – let's call this the *Missing Nutrient Theory*. It stands to reason this would be true for other planets that support highly evolved life as well. A complicated intuition tells me that a species would not evolve to the point of space travel, consciousness, or some other magnificent endpoint if there was a shortage of some nutrient, since that rule seems to be the case with us: our constituent parts are representative of the planet that gave rise to us, wherein those components of which we are comprised exist in enormous excess (specifically carbon, hydrogen, oxygen, and nitrogen). It seems there might be some cosmic rule that determines this, an axiom, call it the **Law of Excess** – that a species could only evolve where there was an enormous excess of nutrients –

whatever elements it needed to not just come into being, but to evolve to some advanced state. There's not just a little bit of life on Earth, there's an incredible EXCESS of it, and the stuff needed to make all this life correspondingly excessive, such that it might *never* be depleted. It seems this would be true for all planets *that harbor highly evolved life.*

> **A species would not evolve to the point of space travel, consciousness, or some other magnificent endpoint if there was a shortage of some nutrient.**

And so it would seem incredibly ridiculous – God playing a joke! – for a species to evolve – to the point of space travel! – at which point some nutrient was used up! Or thereabouts, even within hundreds of years – and that they should get into a spaceship and...go grocery shopping. The indignity! So if aliens came to Earth, it would almost positively NOT be in search of food or other vital resource.

On Earth there are, of course, people starving, but this is due to desert conditions or poverty, not "missing nutrients," not a planet-wide shortage of some vital substance. But even as an analog to the "missing nutrient" problem, starving people – don't develop space travel – they just die. Nor do starving people ask SpaceX to hurry, what other planets might have food! It's an ironic coincidence indeed that some humans are starving while other humans become highly evolved and develop space travel, but it doesn't contradict my theory of the missing nutrient – that aliens wouldn't travel here in search of a missing nutrient or resource. At no point in the history of Africa did widespread starvation and hostile living conditions (due primarily to desert conditions) ignite the development of space travel. And the closer Africa may get to sensing a problem, the closer they will get to extinction, rather than getting closer to *developing space travel.* So it would certainly be an ironic twist if Africa – at any point in the future – develops a space program in order to find a planet that is more conducive to life – by those most affected by starvation. And so it seems this would be true as well for species on other planets that might be impoverished and experiencing a nutrient shortage – that they would not at all be likely to develop space travel – nor, for that matter, develop to any degree of sophistication at all. Nutrient shortage equates to impoverishment and that does not lead to space travel. This is just a convenient example, no larger ethical lesson is intended. So it stands to reason that aliens would not travel this far in search of a missing nutrient or resource – this is just another epistemological fallacy. They therefore wouldn't travel this far for a lesser reason, like *curiosity*. But could there be a greater reason?

Let's look at this idea again, Earth as a giant incubator of life – but not as something disgusting. A planet size garden of Eden. Might some advanced intelligence be exploiting this already, in some way? Such that we needn't worry about colonizing Mars or exoplanets in so-called Gold-ilocks regions? Fear not! As fulfillment of some higher order master plan – godlike in nature – or a survival of the fittest on a galactic scale – is it possible that super intelligent – and beneficent – aliens are, indeed, visiting us...and taking some of us, and other species – our DNA – to plant the seeds of life on other planets, many light-years away?

> Maybe that's how we got here! As part of some endless cycle, with no beginning and no end, just the perpetual continuation of some endless cycle. If we could assume this to be true, we would free ourselves from this burden of getting to Mars.
>
> Just think, man does not need to worry about spreading his DNA– as if our DNA was worthy of such immortality. Whatever higher intelligence that is involved in our creation (as some seem to think) has most certainly taken care of that already and would not leave that impossible burden to us. But it's certainly preposterous to think that human DNA in particular should be immortalized in some way, given our fragility and irrational minds – certainly there is some creature out there more deserving.

MARS WOULD BE THE OBVIOUS CHOICE FOR ALIENS IN SEARCH OF an Earth-like planet. If some alien species was in search of another home planet for itself (the dying-planet cliché), and if they're technically advanced enough to get *here*...it makes sense that they would at least use Mars as a stopping point, if for no other reason than the fact that it would be a shorter trip for them. In fact, Mars would be the obvious choice for any aliens in search of an Earth-like planet. (Which begs the question: why are all the aliens whose planets are dying looking for

Earth-like planets, but let's completely ignore that question for now.) It's not as if aliens searching for another home planet would need to "make contact with humans," which is, after all, nothing but a paranoid delusion embraced and perpetuated by Hollywood, so our *absence* on Mars would not be a problem for them. Using the logic of the Mars advocates, that Mars is habitable and Earth and Mars have at least vaguely similar resources, and the widely accepted belief that what's habitable to us is habitable to aliens (and who am I to challenge popular opinion?), there would be absolutely no reason for aliens looking for a new Earth-like home to settle on Earth instead of Mars. And "looking for a new home" would be the only conceivable reason for any alien to make such a trip – rather than curiously wandering by and then the return-trip. And there's nothing stopping any aliens from doing this! For aliens shopping for a new planet (and if they're already in this quadrant of the universe) everything is on the plus side for Mars. There are no germs on Mars, no angry humans with their philosophies and religions and guns, no other creatures to compete with whatsoever! Mars is so peaceful, and perfect for any self-respecting alien species. Mars is a virtual alien magnet and much more suitable to our beleaguered aliens than Earth! Perhaps we should put up a big detour sign, pointing to Mars, in case this isn't clear to them. But certainly, as I suggested just a few paragraphs back, Earth would be snubbed in favor of Mars and a sign should not be necessary. All of which, of course, seems to be *axiomatic*.

But for practical purposes, aliens may not only reject Earth, but also Mars – for being too close to Earth! – and its germs and religions and philosophies and wars. Any transcendent creature, which is what an alien would be if it could get this far, would want nothing to do with any of these things – and would probably have its own baggage. They might even devise a war machinery to keep *us* away from *them*. Who would have thought that the LAST thing the aliens wanted was to "make contact" with us? This would, of course, be in keeping with the Cosmic Law of Indifference which I proposed earlier. To any transcendent creature about to settle on Mars, the prospects of being visited by humans – *invaded by humans* – may be enough to scare them off to search for a home in another quadrant entirely!

Humans going to Mars is like angels coming to Earth with their wings ripped off.

In the same way that anybody who wants to live on Mars is crazy, anybody born on Mars who wouldn't want to live on Earth would be equally crazy. Let's look at the reverse of my Earth as incubator theory. Let's say humans were brought here many years ago from an Earth-like planet where our progenitors had wings but they were removed before they were brought here (because of Earth's strong gravity). Wouldn't you be resentful that this ability had been taken from you, and do anything to get it back, even return to that planet to get your wings back? Our constant preoccupation with flight and flying suggests such an intuition. This same logic would be true for people born on Mars looking at Earth, where, even though we don't have wings, we're not dragging around a ball and chain all day long (oxygen tanks, pressure suits, gravity boots) or constantly aware of simply staying alive in the unending struggle with a planet that's constantly trying to kill us. Would be true even if Mars was more like Earth – would – yet again! – be true for an Earth scorched by apocalypse and ruled by demons! THEREFORE, humans going to Mars is like angels coming to Earth with their wings ripped off; future Martians will be like angels with their wings ripped off – crippled, subverted. Alien.

Let's complete this blazing analogy: compared to Mars, Earth is Heaven. And who – you? – not me! – would want to leave Heaven?

THE FIVE F'S

The others – aliens – about whom many fantasize are presumably more intelligent than us, and have answers. But eventually we will discover the answers ourselves, which should obviate the need for us to continue searching for alien intelligence. Given our advancing knowledge, at some point we would be superior even to alien species, even those smart enough (or dumb enough) to contact us, at which point they might serve no purpose other than to be killed for food or amusement – which is how we treat all other earthly species now – and, sometimes, each other – which leads to my *Theory of the 5 F's*. Or should I say religion – the one *true* religion. Whatever intelligent alien species that may have visited us in the past has been wise to be discreet about it. Humans have freely killed all other species on Earth, either for food, fur, out of fear, or for fun, or the many we *fence* in and eventually kill. The 5 F's. This describes our relationship with all other species on the planet – including, at times, other humans. Do we really think we'd treat the next species we discover or encounter any differently?

The 5 F's– are you in shock? It's so funny, or sad, how mankind maintains this innocent idea – some of us – that when the next creature walks down the street, or lands out on the lawn, we will treat it differently. Become friends. We will completely disregard our religion of the 5 F's and, instead, *share the secrets of the universe*. The aliens will not just speak English, but the vernacular. *We'll get down with them. Mutual cosmic destiny.*

As if there could be more harmony between us and aliens than between us and each other.

Nothing will discourage our welcoming arms – we will immediately recognize and seize the moment, like long lost friends...all of which will literally ignite a new age of medical breakthroughs, high employment (cheers), galaxy-wide serenity (more cheers), and space tourism.

Because the reason we exist is to fly around at the speed of light, like photons, and the aliens are coming to take us there.

And our alien friends won't be 20 feet tall, nor 2 inches tall, nor 2 microns tall – they will be a spiffy, homo sapiens compatible 6 feet tall. Like in the cave drawings. And we will mate with them and make cute babies – if we haven't already.

It's interesting that we suspect that any creature that would land in our back yard would know anything about the secrets of the universe – or know something about *cancer*. Such a creature would more likely have questions than answers – which might work against us regarding any ideas that they would immediately hand over *the cure*. We landed on the Moon full of questions – that's the only paradigm we know that actually exists. It's the nature of every space traveler we ever knew to have questions rather than answers – based on our one and only example: us. So I suspect this would be the case for space travelers coming here, that they would have more questions than answers. And that any idea that they would be spilling their guts on the secrets of the universe...is completely unrealistic.

BUT FINALLY, THERE ARE PROBABLY NO OTHER CREATURES LIKE us in the entire universe – not even vaguely. Certainly not something that we would treat as an equal. Even on Earth, the closest thing to us, the apes and monkeys, are really quite different. To think that something in another star system would be more like us than an ape, and might come here and have a natural life on Earth, is simply not reasonable, yet this would have to be the case if any version of "contact" that has ever been imagined were to be realized. So, aside from any creatures out there that are *less* like us than an ape – which would be irrelevant to our search for other creatures "like us" and with whom "making contact" is not an issue,

It would seem to be an epistemological fallacy to think that something that is more like us than an ape would exist in some other star system – but not here.

As with a jigsaw puzzle, if a piece is missing, if it exists at all, it is bound to be somewhere in the same "place" as the other pieces – to think otherwise is an epistemological fallacy. Or, to put another way, on Earth we have creatures with two legs, creatures with four legs, creatures with six legs, and creatures with eight legs (and creatures with more than eight but let's ignore those for now). This suggests a systematic progression. Would it make sense if there were no creatures on Earth with six legs, everything else being the same? If that were the case, one might intuit that something was missing. And if you pursued this intuition, suspecting that creatures with six legs must surely exist, wouldn't you expect to find such creatures somewhere in the same world as the 2, 4, and 8-legged creatures? Or would you think, surely the missing 6-legged creatures must be living on another planet? Of course you wouldn't think this!

Then why do we think that creatures that are more like us than an ape would exist on another planet in another star system! It seems axiomatic that anything more like us than an ape...would exist in our world – Earth – and not another.

So, in terms of "creatures like us," it seems a strong case can be made that *we're it.* And rest assured, none of the sci-fi fantasies that have ever been written or imagined, that involve alien invasions or visitations, will ever come true. There may be life on other planets – most everyone is in agreement with this and it's really not such a provocative idea anymore – but not "creatures like us."

50

Speed of Light Travel

||||||||||||||||||||||||||||||||||||||||||||||||||||||||||||||||||||||||||||||||||||||||||||||||||||||||||||||||||||||||||||||||||||||||||||||

This axiom – that there is nothing we need that is so far away that we need to travel at the speed of light to get it – will become anathema to fans of "speed of light travel," if not the entire space industry or the Mars advocates. This axiom could single-handedly undermine, or determine, the entire future of the space industry.

We may have reached our max "low atmosphere" air speed – the SST's are history.

IT IS AXIOMATIC THAT THERE IS NOTHING WE NEED THAT IS SO FAR away that we need to travel at the speed of light (or even a fraction of this) to get it. And how cool is that! Everything we need, absolutely positively everything, is *within arms reach*, relatively speaking. This is, of course, how every species on Earth evolved, as a function of the environment in which it exists – everything it needs to not just survive but thrive is right there in its natural habitat. Of course getting food may involve some walking or flying (if you're a bird), but these forms of *locomotion* can be taken for granted when I say that everything we need is within arms reach, you don't need to take the expression literally. Since my point is to say how complete, and convenient, our lives already are, compared to a life that would require us to travel to some other solar system as a matter of routine – as might be envisioned by fans of Star Trek who think this should be inevitable, but for which the opposite seems to be what is inevitable. In either case,

Whatever effort is being diverted toward a Star Trek existence – and which is certainly in vain – it makes sense that an even greater effort would be made to preserve Earth and ourselves with it, no matter what spectacular inventions that may entail.

So a speed of light craft will probably never exist if only because it's impossible to justify one – you would need a pretty good reason to justify such a preposterous thing. Curiosity alone would not be enough – or even the presumption to *save humanity*. Necessity, not curiosity, is the mother of invention, and it's ridiculous to think that curiosity could ever be the driving force to something as spectacular as speed of light travel. It's easy enough to justify travelling to the next town over, or another state or country, but another star system?

Even aside from these considerations, man will never travel at the speed of light, it's impossible for several reasons. The scientific community does not debate this – but at the same time scientists have taken to studying exoplanets, inspired more by science-fiction than practicality because otherwise they may as well close up shop and go home – that's the state of astronomy today.

But for the sake of discussion: since light-speed travel could only be relevant in travelling to another solar system, it might be discarded for that reason alone, since even travelling at the speed of light, the closest solar system is over 4 years away. Light-speed travel really wouldn't be useful for travelling within our own solar system because, basically, the other planets in our solar system are too close to us. The issue is the time it would take for a speed of light craft to accelerate and decelerate. Any craft carrying human passengers would need to accelerate and decelerate slowly, so accelerating to and decelerating from the speed of light without killing the passengers from extreme G-forces would take a long time – such that by the time a speed of light craft reached light-speed, it would have already left our solar system! Which is true for even a fraction of the speed of light! In which case any technology that would allow speed of light travel would be pointless and a complete waste of time to pursue.

If you're wondering what distance might be practically travelled by a speed of light craft – or what destination might be at such a distance – the answer is: none.

Any theoretical destination would be somewhere between the edge of our solar system and the next one over – but that's just empty space! That's right, there are no destinations in the space between the edge of our solar system and the next one – because there are no things in that space, just vast nothingness. Anything *further* than that would exceed practical limitations, being on the order of one or many lifespans, regarding solar systems beyond the one or two that are closest to us, though this particular statement could lead to heated argumentation among those whose minds are warped by Star Trek and other fantasies, in defiance of the practical burdens of being human.

You certainly couldn't use speed of light travel on Earth itself, as if one might travel from New York to Los Angeles at the speed of light, because the G-forces during acceleration and deceleration would kill all the passengers in such a craft, probably even melt or otherwise destroy them in the process. They'd be crushed to the thickness of a few atoms. They'd be dead before they left the runway. Plus you have other issues of practicality, such as why would you need to go from New York to Los Angeles at the speed of light? Speed of light travel could only possibly be useful for travelling between stars, and borne of great necessity, given the "infinite" resources that might be required to do this, but there is no such necessity because everything we need is "within arm's reach," so such a thing would be frivolous at best. A fun science-fiction idea that deserves to be nothing else.

LET'S IMAGINE TRAVELING AT THE SPEED OF LIGHT, FROM NEW York to Los Angeles. The entire trip would be spent either accelerating or decelerating – the craft would never reach a steady cruising speed! And the sensation of speed would be constant, like when you're in an airplane taking off and it's going faster and faster, and everyone's in a state of anticipation, with their seat belts fastened, everyone in a state of suspense, pushed back into their seats. We don't feel speed – we feel change in speed, and the speed is constantly changing as the craft accelerates, faster and faster, toward the speed of light, but never actually reaching this speed because by the time you got just halfway to your destination – and your craft is still accelerating – it would need to start decelerating! You would never get to a sustained light-speed cruising speed, at which point you could relax and become unaware of your intense speed, but instead start the gradual deceleration, which of course you would feel, might even be uncomfortable, and need to be strapped in, in a state of suspense and anticipation as with the entire first half of the trip, the deceleration as controlled as the acceleration so as not to disrupt your comfort, or cause adverse health effects, or flatten you to the thickness of an atom. Not that this less-than-light-speed trip would be technically impossible, but no part of it would be at – or near – the speed of light. So you see, speed of light travel would not be feasible for travel between locations on Earth – only for travel to other stars – but even then it would take years – lifetimes – of travel time! How could this be a worthy ambition? It doesn't make sense. Anyway, most scientists agree that speed of light travel is impossible even in theory.

THE PROBLEM WITH SPEED OF LIGHT TRAVEL

Imagine a space craft with a camera attached sending back the image of what it sees, and the space craft is travelling at the speed of light away from you. Each second that passes, it would take a second longer to receive the image (via the electronic signal it's transmitting), such that the liveliness of a "live image" would decrease, like a movie that's going slower and slower at a speed that decreases by 1 second per second, i.e., it's decelerating. At some point – after only a few seconds – the image would become frozen, or appear that way – so the idea of monitoring a space craft in such a way would be pointless, and whoever was on such a craft would be on their own (unlike today's aircraft or rockets that receive signals from the ground while in flight). Or let's say the craft is sending back a single image every second. When the craft is 1 light-second away, it takes 1 second for us to receive the image (ignoring the time it would take for a speed of light craft to accelerate). Likewise for 2 light-seconds, 3 light-seconds, etc. So it would take 1 second to receive the first image, 2 more seconds for the second image, 3 more seconds for the third image, and so on. Even though the camera on the craft is taking a picture every second, you receive the first picture on the count of 2, the next on the count of 4, the next on the count of 7, and so on, so that in the first 60 seconds you receive only 10 pictures, not 60 (actually, you would receive the 10th picture on the count of 56). So it might seem that the camera was broken and not taking pictures every second. Or that the camera was only 10 seconds away even after 60 seconds...but because it's travelling at the speed of light it's actually 60 seconds away, it just takes longer for each successive picture to travel back to you. Confused?

The point is, such issues concerning relativity (Einstein's Special Theory of Relativity, to be exact) would make it difficult to develop and test speed of light craft, or even fractions thereof. But aside from the relativity issue, in developing speed of light craft, or fractions thereof, we wouldn't be able to observe the thing we're testing, in the way we're accustomed, which would violate one of the primary tenets of science: observation. We couldn't simply confine our experiments to a particle accelerator, which would be convenient, where everything just goes round and round. This helps to explain how complicated things are starting out, just to study photons and other speed of light particles.

So how do you experiment with much larger objects, like space craft? We might launch a near light-speed craft – and not know what happened to it! If a near light-speed craft existed, we couldn't keep track of it, follow along on its journeys – its discoveries would be known only by its crew (assuming it had a crew). And its journeys would probably not match the excitement or thrill suggested by science-fiction. There are, of course, intrinsic limits to how much we can learn in this area, where all kinds of uncertainty principles are in effect – we are getting closer and closer to that point. In this arena, for progress to continue as it has in the recent past, or to accelerate, we would need to make some leaps of faith, which are, basically, a scientific no-no, and which also infringe upon our standards of safety – which have become increasingly less risk-tolerant. That's one reason nothing has happened since the Moon. Or, besides leaps of faith, we'd have to redefine science itself – expand upon the utterly sacrosanct *Scientific Method*. Or, whatever it is that progress entails may be beyond our comprehension, involving unknown and unknowable dimensions – and terrors.

Despite all this, some engineers like to imagine – in fact they are paid to imagine! – that some day we will travel among the stars, and to that end they sit around and invent theories and create computer models in the hopes that this will eventually be possible. There have been a few scale models and prototypes that have received attention – from others warped by Star Trek and other fantasies – but it's mostly just calculations and artistic renderings with an occasional movie quality animation that can be used to solicit funding for some commercial enterprise, for self-aggrandizement, or to otherwise inspire others to join their ranks.

There are at least four types of ultra high-speed propulsion systems in various stages of design – still all very theoretical, some very outrageous – that might, according to their proponents, power spaceships to the stars. But which would take decades or centuries just getting to the closer ones – decades or centuries! Why not have a single, fool-proof Plan A, that way you don't need to waste time on a Plan B, Plan C, and onward through the alphabet. Evidently every plan has reached the point where too many questions can't be answered, so that drawing is put aside and another begun...and so on through the alphabet. We're just not smart enough to complete any one plan. If man had an IQ of 130 or so – 200 – rather than the stubborn 100 (and falling), there would be someone who could complete the drawings! Not play the piano or become a neurosurgeon, but finish the drawings! We'll all just have to wait for that race of geniuses to be born, to finish the drawings...so we can finally free ourselves of this misbegotten Garden of Eden.

51

Asteroid Mining

And what are we going to do with the gold brought back from asteroids – bury it deep underground with the other gold reserves, like spent plutonium? And reserved for what?

I'm about to say something that will upset some of you, and may even ruin your day: Gold has no real practical value and is more or less useless otherwise, other than for making coins but its role in coinage has, since the early 1900's, been replaced by other metals. If you bury something underground, as one might bury nuclear waste – which is what we do with much of our gold...that means you either don't want it or don't need it. As opposed to something you need and that might be within arms reach, like the food in your refrigerator and I don't think I'm being overly simplistic. Besides gold and spent plutonium, what else do we bury underground? Of course you may invest in it but that's based on speculated value, not real value. You might also invest in copper – something that has real practical value – but no one would travel to an asteroid for this.

Its role is existential, insofar as it *fills a slot in the periodic table*. The elements above, below, and on either side of it in the periodic table requires that it exist – it has to exist, because as you go through the periodic table, from atomic number 1 through 118, each element has one more proton in the nucleus of its atoms, from 1 to 118, and gold just happens to be the element that has 79 protons. The series is complete, nothing is missing – every element, with a number of protons corresponding to a number in this series (the atomic number) must exist, it can't be any other way (whether it occurs naturally or is man-made). And therefore gold exists.

But the value of gold is largely mythical so isn't it time we lose our El Dorado fantasies? This is another one of those *atavistic* human needs – which puts asteroid mining at the top of the absurd scale. We don't need or even use the gold we already have. *It's the only element about which people have feelings.* It's not consumed the way other raw materials are consumed – to be chemically altered or destroyed in the process – and is easy to recycle. All the gold that was ever mined still exists, and there's lots more right here on Earth. Currently there are over 80,000 gold miners in South Africa alone. Does this suggest to you that there is a shortage of gold? But it's an arbitrary reference. The value we place in gold, and the universality of this, only demonstrates how irrational we are as a species, and how universal this irrationality is! It's actually one of the least useful of the more widely known elements – why else would we sequester it in vaults and personal hoards, where it clearly serves no utilitarian purpose? So the idea that anyone would think they should rendezvous with an asteroid – an asteroid! – looking for gold – gold! – has got to be the stupidest idea a person could have. To think this idea would come from a scientist, or be supported by science...is too dumb for words. It's just more of that venture capitalist brand of *idiot savantism*.

If all the gold on Earth suddenly disappeared, we would be none the worse off (except perhaps for very tiny, tiny trace amounts found in our bodies, which evidently make their way from natural amounts found in the Earth but which would obviously not be affected by gold retrieved from asteroids). Other than these trace amounts, which are unaffected by human activity, the only thing gold is good for is hoarding and tooth fillings. The reserves in Fort Knox and another in a vault in Manhattan prove this, plus the 75% of all gold that is used for jewelry, and that stored in privately owned hoards. It's disgustingly overrated. There's increasing demand for it? What, by families in India who hoard it for their dowries? (According to CBS's long running and award winning weekly investigative TV show *60 Minutes* on February 12, 2012). These Indian hoards account for over 30 percent of the worldwide market for the metal. Who knew! Imagine rockets launched from India in search of gold! How badly does anyone need...more gold jewelry! To think that anyone would go to the colossal trouble of "mining an asteroid" – even if it was possible – proves that some venture capitalists are no smarter than anyone else. As if mining for gold *on Earth* wasn't dangerous and deadly enough. And when I say "even if it was possible," it seems there should be no need to make the argument, because why would anyone need to wonder if such a thing was possible!

As with other rare-earth metals, it's enormously unwise to become dependent or more dependent upon something that doesn't already exist in abundance on our own planet – like an addict looking for something else to become addicted to – a terrible, terrible idea. If something exists in trace amounts, you don't contrive to look for a million and one uses for it, that's just plain stupid.

> **What we need to do is round up these amoral venture capitalist hawks to deposit on the nearest asteroid as it slingshots around Earth toward the Sun – and for them to take all the gold we now have on Earth...with them!**

From before snake oil to asteroid mining and beyond, this kind of nonsense may never cease, and is in its own category of ridiculous.

One can make a corollary to the Law of Excess, which I propose in the *Aliens* chapter, such that elements *not* essential to our survival would be rare. In which case it's not ironic that there is so little gold, rather, it is only too reasonable, as if our existence and evolution were somehow dependent on its *scarcity*.

The so-called "market value" of gold should have no influence whatsoever in the pursuit of extraterrestrial sources of gold, nor even to investigate such opportunities, since its market value is both unreliable – and ultimately meaningless. Gold is not vital to our existence. The role that gold plays in our culture is accidental and to our *misfortune*. If gold or any of the other rare elements could make us immortal...that would be different! – but it represents nothing but bourgeois materialism. Gold will eventually lose its current high value – or it may increase – it doesn't matter! Venture capitalist gold diggers and adventurers will look silly either way – such a far out venture could never be win-win – how crazy! Even if an asteroid were solid gold it wouldn't be worth it – the traditional value of gold is due to its scarcity – if gold became plentiful its value would plummet, so the idea of mining asteroids as a wealth building strategy could be attractive only to an impoverished gringo. No smart money is riding on the idea that there are mineral riches either on Mars or on asteroids – still, some will wear their stupidity like a badge of honor. It makes more sense to count cards or play the lottery.

Our feelings toward gold demonstrate how trapped we are in our earth-minds. On Mars there would be a completely different value system, but evidently this is difficult even to anticipate by the Mars advocates. So let me spell it out for you: on Mars, water and oxygen would be the precious commodities, not gold. On Mars, man will not be mining for gold because on Mars, man will not be thinking of Earth – he will be too busy staying alive and being a Martian. What a petty, bourgeois, and earthly hang-up, to think that men on Mars (or asteroids) would be mining for gold!

52

Thoughts & Reflections

> Humans are part of the nature of Earth, the child of Earth. Could we ever be part of the nature of Mars – or always the step-child? Would we always just be visitors, unable to escape the singular imperative that we are Earthlings? That would keep us from fully embracing Mars, and Mars from fully embracing us.

MANY OF THE REASONS MAN CAN NEVER LIVE ON MARS ARE NOT as shocking or stark – or obvious – or deadly – as the lack of oxygen, food, and water – or worthy of their own chapter – but are compelling nonetheless. Whether the deal breaking implications of the following considerations are immediate or far-reaching, it is impossible to think that Mars could ever be a rational choice and one big question remains: what wonderful things are the Mars advocates imagining are on Mars that could possibly compensate for its countless deadly or miserable shortcomings?

◆ ◆ ◆ ◆ ◆

As I stated in the Introduction but which will benefit from a certain redundancy, our continued fascination with Mars demonstrates, if nothing else, how we hang onto both the past, and to antiquated visions of the future. That's right – our vision of man on Mars is not futuristic, but antiquated.

> **Mars – is passé. Manned space flight was a phase that has run its course, and henceforward is a complete waste of time – or a frivolous diversion of the idle rich, which it is quickly becoming.**

◆ ◆ ◆ ◆ ◆

What else can one do with Mars but send up more rovers! It demonstrates several things:

1 **The natural limits of human ingenuity**. I mean, how many "big ideas" can you expect? Even the most remarkable advances in science don't occur in leaps and bounds, rather in tiny, incremental steps – which *exploit*, rather than *defy*, the laws of nature. A colony on Mars might be a spectacular, enormous leap but is not supported by existing science and defies the laws of nature. You can bet that, even taking incremental steps, we'd never get to that point, given our history of stumbling through space (e.g. the countless space programs that either failed or were decommissioned such as the Shuttle program and the Concorde SST). Additionally, the state of the art of rocket science and the fantastical "terraforming" do not and may never allow us to travel to or colonize Mars.

2 **The fact that man is not capable of living on Mars**. Bear in mind that if man ever gets to Mars he would need to be entirely, 100% self-sufficient, since he wouldn't be able to call Earth and say hey, what do I do next. On Mars, this means that life would never exceed a bare bones, life and death struggle. We already know that Martian dust can't grow plants, due to a lack of the necessary minerals and gases like nitrogen, among all the other things already described. So things really do NOT look promising.

3 **It's just another day at the office.** Your day is mostly a repeat of the previous. Same with the guys at NASA, and everyone else. They will keep doing what they know! Human life is, ultimately, based on repetition, the status quo.

4 **Other planetary bodies are just too far away!** A single project must be conducted during a tiny portion of a single lifetime for it to have meaning to anyone. The fact is, people want to see the results of their work. Face it, space-time is BIG. We can't overcome this reality.

◆ ◆ ◆ ◆ ◆

The rovers are designed to detect obvious life forms or signs of life. Living creatures are comprised of organic compounds, but organic molecules alone (in case they're detected by a rover) do not comprise life, nor are they a substitute for a creature that's alive now, or its corpse, or its fossilized remains. Other than such direct evidence it's impossible to know if or when life ever existed on Mars, no matter how many dry riverbeds crisscross the surface. One can't simply extrapolate from the discovery of "organic molecules" whether life may have existed on Mars at some point in the past. But the point seems to be moot since the Mars advocates are not relying on such discoveries to fulfill their destiny of colonizing Mars.

◆ ◆ ◆ ◆ ◆

It is not an expression of the human spirit, but a perversion of it, the most extreme kind of perversion, that we should leave Earth.

◆ ◆ ◆ ◆ ◆

Astronauts are obsessed with taking pictures of Earth, their numbers can't be counted – some so beautiful you could cry. Doesn't that tell you something? What does it mean that astronauts in space are so in love with Earth? They're H-O-M-E-S-I-C-K. It seems inevitable that man on Mars will be the same.

Are astronauts nothing but shutterbugs? How many more pictures do we need? And every picture is from the same perspective: inside the ISS.

According to a new revelation (National Geographic, Jan 2013, p. 20) the space shuttle astronauts pined for small reminders of life back home – for such things as the music used for wake-up calls. Even after all the training and preparation. This does not bode well for a mission to Mars that is one-way. What will prevent the unavoidable psychological toll from crippling the chances for a successful colony?

◆ ◆ ◆ ◆ ◆

There will be no outdoor concerts on Mars – since there is (practically) no air on Mars and the transmission of sound waves requires air and humans on Mars would be wearing air-tight helmets anyway. TGIF! Or maybe not.

◆ ◆ ◆ ◆ ◆

From man to monster

In many ways man is still the caveman he was thousands of years ago, a bit more intelligent, with better language skills (because we spend years developing this in school) and a lot more tools. But man basically has the same "I want food and sex" brain – or the more modern "I want food and sex and TV." That's our nature and that won't change. Creation has imprinted us to be Earthlings and that seems to be the grand design. Going to Mars or any other planet disrupts this master plan. In either case, people don't want to live on Mars, it's just not something anyone really thinks about, nor does anyone want to be bothered with the question – even if faced with alternatives that are apocalyptic.

What makes us different from all other living creatures is our consciousness, our self-awareness, and the degree to which this allows us to control not just our lives but the destiny of our species. But our ingenuity and civilized behaviors mask our true animal nature and the savage beast within all of us. This would not be the case on Mars, where man's basic nature would be more unavoidable and stark, emerging rather quickly and with shocking results. Mars would bring out the monster in us. Like cannibalism. Even where there might still be a months long supply of "food in a tube," humans may be forced to eat each other and not necessarily as some zombie-esque nightmare. Human flesh might become a preferred source of protein, if only as a more extreme form of recycling, where nothing – nothing! – is wasted. Why wait when you're hungry now – got to eat! Imagine people – people! – being raised...as a food source! How can anyone think of Mars without imagining this or other nightmares? Who can know how scary it could be on Mars, to actually be there, where there would be no relief, or hope for relief.

◆ ◆ ◆ ◆ ◆

Even on Earth, a relative utopia, basic infrastructure is incomplete, affordable only to the wealthier nations. Furthermore, we need it to be between 70 and 72 degrees, and create machinery to maintain this finely tuned environment for ourselves at great personal expense but also at great cost to the environment and the planet. Even in highly developed countries like America some die during periods of extreme weather – not from direct exposure to the elements but *in their houses*. People who can't afford heat or air conditioning. And of course nobody cares but which could be easily prevented. Obviously this precarious dynamic would only be exaggerated and more complicated – unmanageable – on Mars. Yes, people die in the relative balminess of highly developed countries such as ours, with or without 911. *How could things be better on Mars?*

◆ ◆ ◆ ◆ ◆

The sobering truth is, there is no analogy to going to and building a permanent settlement on Mars. A settlement on Mars would change our perspective of everything on Earth, it would change everything. Even our language would change. We would think differently. In colonizing Mars a completely new dimension would be added to the decision of where people might live, which would be as true for anyone living on Mars as those on Earth. As if anyone might wake up one morning and think, "I'm moving to Mars." A colony on Mars would change the calendar – we would start using A.M. for Anno Mars, in the year of Mars. But these things will not happen. We may walk upon the surface, but no permanent settlement will be established.

Look how much fanfare there was regarding colonizing the Moon, throughout the 1960's and 70's, and afterward, but which amounted to nothing. What does it tell you that Neil Armstrong and Buzz Aldrin spent all of 2.5 hours on the Moon and left as fast as they arrived, as if there might be a monster lurking in the shadows. What's your hurry, guys? You prepared all those years, NASA spent vast fortunes inventing delicious, prepackaged foods. Apparently all for nothing! After all, it's not as if there would be a reason to go to the Moon – what could one possibly do on the Moon – besides moon walking – "Look, I'm bouncing!" Of course we never really intended to colonize the Moon – we fabricated a "been there done that" moment while providing an entire generation fodder for popular science magazines and jazzing up our "national pride." And that spacesuit was impossible to move around in, bouncing around...my hunch is that it will be basically the same with Mars. Mars is the new Moon. It's a matter of pushing the envelope for its own sake, with no real practical benefits, other than of course all the make-work jobs that are created and for which we will praise and anoint ourselves with holy oil. And of course a few other tools and toys which will amaze and *make us proud*. As is so often the case, *anticipation exceeds the realization*. This will be as true for Mars as it was for the Moon – and there's nothing on the Moon but moon dust.

53

The Self-Conscious Universe

To the extent that we are aware of the universe, the universe is aware of itself, so when you look up at the stars, that is really the universe gazing upon itself. That is our significance – simply to be. All we need to do, is exist, and to revel in that. If we can do this, the universe will be forever.

THE FINAL ANALYSIS SEEMS UNAVOIDABLE: MARS IS NOT SUITABLE for colonizing, any size colony could not be self-sustaining, there's no point in colonizing, there will never be 80,000 humans on Mars, nor 8,000, nor 800. To get just four guys to Mars, scout around for a few days, then all safely return to Earth...would be a major miracle – a triumphant convergence of technologies – no matter how far into the future. Such a thing is not inevitable and it's pure hubris for anyone to think it is. But the stark reality is that even this is *impossible* in this day and age. Still, if humans ever get to Mars it will quickly become a case of been there/done that and like the Moon will become a footnote in history.

For the sake of placating the Mars advocates, however, I'd like to end this on an encouraging note – after all I'm not against outer space exploration in the general sense or a pointless iconoclast. There will be fantastic discoveries, made increasingly easier by both Earth and space-based telescopes – which is certainly convenient and will not cost an arm and a leg. This is exciting and will advance our frontiers. But we will always be in the state where we are now: enthralled. Perpetually enthralled by possibility, while forever trapped within the reality of our human existence – and the reality of Mars' obvious inhospitableness… in addition to our own inability to see beyond the present with more objectivity and wisdom.

Some day in the far, far future we may chance upon an exoplanet that has physical characteristics similar to – or better than – Earth. Some day we may develop the type of propulsion necessary to get to another solar system but which we can't possibly imagine now. Both of these two milestones may be inevitable. They may be centuries apart. But we will need these, and other, breakthroughs before we can "travel among the stars" with any degree of purpose and meaning – while retaining our humanity.

OUR AWARENESS OF OURSELVES, AND THE UNIVERSE, MEANS THE universe is aware of itself, or self-conscious. This is mind-blowing! Strip away our clothing, our routines, our devices, our vices, our virtues, and what's left? We're not separate from the universe, we're part of it – our atoms, our molecules. The universe is literally aware of itself to the extent that we're aware of the universe, when we gaze out upon it. When you gaze upon the stars in the Milky Way, and other galaxies, that's the universe gazing upon itself! The universe evolved to the point of self-awareness! And each one of us makes the universe that much more alive. Fate may bind us to Earth forever, but at least we, the eyes of the universe, can dream and gaze upon the stars in wonder, and give consciousness to the universe – indeed, it is that which may be why we exist at all.

When the last human is gone, if that ever happens, the universe will, in a very real sense, be dead – except for other sentient beings that may be on other planets staring in wonder as we do. So that is our significance – simply to be. On Earth, or Mars, or anywhere, it makes no difference. The virtues don't matter, nor do the vices. Good, Evil, whatever. All we need to do, is exist, and to revel in that. If we can do this, the universe will be forever.

In the meantime, Earth allows a freedom that no other planet can – the freedom to be Human. But the future is inevitable, so lift your glasses to the future: where the laws of nature, and the limits of our imagination, shall meet.

Index

About the Author

N.B. McKinney has a B.S. in Biochemistry, a B.S. in Computer Science, and a wide-ranging interest in science and technology. He can be reached at JunkyardMarsBook@gmail.com.

www.ingramcontent.com/pod-product-compliance
Lightning Source LLC
Chambersburg PA
CBHW051439170526
45166CB00001B/48